— Mike
Thanks for your support and being part of this story.
John D. Butler, Ph.D.

Who's the New Kid in Chemistry?

Exploring Uncharted Waters

John D. Butler
With Jessica M. Waters

— Best Wishes

Jessica Waters

12-18-13

UNIVERSITY PRESS OF AMERICA,® INC.
Lanham • Boulder • New York • Toronto • Plymouth, UK

University Press of America,® Inc.
4501 Forbes Boulevard
Suite 200
Lanham, Maryland 20706
UPA Acquisitions Department (301) 459-3366

10 Thornbury Road
Plymouth PL6 7PP
United Kingdom

Printed in the United States of America
British Library Cataloging in Publication Information Available

Library of Congress Control Number: 2013950072
ISBN: 978-0-7618-6229-1 (clothbound : alk. paper)
eISBN: 978-0-7618-6230-7

∞™ The paper used in this publication meets the minimum requirements of American National Standard for Information Sciences—Permanence of Paper for Printed Library Materials, ANSI Z39.48-1992

This book is dedicated to the students in Mrs. Waters' Chemistry Block 1 class
Beacon Charter High School
Spring, 2011

Amanda, Brian, Brittany, Hector, Jessica G., Jessica P., Julia, Kathryn,
Lindsay, Manar, Mary, Nicole, Rosabel, Ruby, Sarah, Sima, Siraj

A new look at teaching, if there is to be one, seems to require us to move up close to the phenomena of the teacher's world. But such a move, though long overdue, is just the beginning.

—Philip W. Jackson, *Life in Classrooms*

Contents

List of Figures

List of Tables

Preface

Not too long ago, I was asked to make a presentation during one of our school's Professional Days. These are the days set aside for teacher training, days on which students are excused from attending school. More often than not, I am asked to speak on a topic or topics that have failed to make anyone's *bucket list*. My presentation, selected by the principal, was on using data to drive instruction (I am the math department chair and data analyst for the school). Anticipating that this would be problematic with the predominantly math-phobic faculty, the principal scheduled my presentation to begin at 10:00 a.m. and paid for an outside warm-up act to my talk—a counselor to speak on suicide prevention from 8:00–10:00. The traumatic realization that my administration felt a need to provide such a service to my colleagues before I spoke was sufficient to cause me to seek therapy in the subsequent weeks! However, after several failed sessions, I decided to write more than speak; hence, this book.

Since the preface of a work is the place where the author generally outlines his or her reasons for undertaking the work and describes the method of research used, I here offer the following.

The year 2001 was not the best year for many Americans. Admittedly, for me it also began a bit roughly: I had to have major heart surgery. During the early days of my recovery, my cardiologist recommended that perhaps I consider making a change to a career that would be less stressful (I had been an engineer for 22 years). I took his advice and became a math teacher in an urban high school in downtown Providence, Rhode Island in December of 2001. I enthusiastically welcomed the opportunity to teach. However, after just a few days in this inner-city school, I quickly realized that when the bell rang for the start of class, not every student was sitting in their seat, book opened, writing implement in hand, wide-eyed and with the sort of body language and facial expression that connoted an appetite for learning mathematics. Since my very first days in the classroom, my focus has been on understanding student engagement and best practices for teachers.

Early in the period of the literature review required for preparation for the PhD dissertation, I came across a book that truly captured my interest and eventually became a major influence on my journey toward becoming an educational researcher—*Life in Classrooms* (Jackson [1968] 1990). Author Philip W. Jackson had spent two years visiting elementary school classrooms in the early 1960's. In his preface to the work, he states: "Its aim is neither to damn schools nor to praise them, nor even, necessarily, to change them. Rather, the goal is simply to arouse the reader's interest and possibly to awaken his concern over aspects of school life that seem to be receiving less attention than they deserve" (Jackson 1990, xxi).

I felt an immediate connection with this work. In the new introduction of the reprinted edition, Jackson tells the story of how his book came about. He was pursuing a career in educational psychology during the late fifties and early sixties. As he explains, this was a time ". . . when educational psychologists were principally test-givers rather than observers of the real world. Moreover, if they ever did observe anything, they did so with a stopwatch and a carefully designed observation schedule in hand" (p. ix).

Early in his academic career, Jackson was heavily entrenched in data-intensive statistical manipulations. The main source for research on pedagogical effectiveness at that time was pencil-and-paper tests. He was becoming dissatisfied with the direction his work was taking him in the profession when he happened to attend a seminar run by a group of social anthropologists. Jackson explains that he knew nothing about anthropology but quickly became fascinated with the many methods of investigation employed in this field, in particular, those of George Schaller, author of *The Mountain Gorilla* (University of Chicago Press 1963). Jackson states: "That methodology versatility contrasted so markedly and so impressively with my own prior reliance on tests and questionnaires that I vowed to imitate it as best I could in my own work" (Jackson 1990, xii). Subsequently, Jackson spent two years sitting in the back of elementary classrooms observing the goings on. Reflecting upon his original work, he states that collecting data to answer preconceived questions, searching for a problem, or exploring the field, did not describe what he was trying to do. He saw his role more as observing and interpreting classroom interaction so as to provide his readers a fresh view.

> In the fields of observation chance favours only the prepared mind. —Louis Pasteur (Lecture, University of Lille, December 7, 1854)

On Saturday, August 16, 2008, I received an email from Professor Louis J. Heifetz, my major professor in my doctoral program. After thirty-five years of teaching, that included five years at Yale, twenty years at Syracuse, and, most recently, ten at the University of Rhode Island, Lou had announced his retirement in October of 2007. He had, however, agreed to work with me until I completed my doctorate. The university had allowed him to keep his office until the space was needed and that time had come. His email read:

> In cleaning out my office, I have tried to be hard-nosed about not keeping anything that I'm not likely to make regular use of. As a result, on the floor beneath the mailboxes (near the elevators on Chafee's seventh floor), I have put many books and journals (lots of them quite current) for folks to take. If you have a chance, you might want to take a look at them before students return. Please take anything that interests you. I'll be in touch. Regards, —Lou.

Realizing that students would be arriving on campus in a few days for the start of the fall semester, it became imperative that I travel to the Kingston, Rhode Island campus as soon as possible. Within one hour of receiving Lou's email, I was pushing the elevator button for the seventh floor in Chafee Hall. I found two cardboard boxes stuffed under the faculty's mailboxes overflowing with miscellaneous books and grabbed anything related to math, statistics, and educational research—topics of interest to me. No compassion for other potential readers! Just the greed of someone too often labeled as a hoarder! I was reminded of Pasteur's "the prepared mind" as I culled the materials together. I stuffed my empty leather book bag so full that the zipper no longer meshed and then scurried off into the evening shadows.

Back in my apartment, I emptied it onto the floor and began skimming through my newly acquired books with the intention of prioritizing them. One of the discarded titles that I had grabbed was entitled *Issues in Participant Observation: A Text and Reader* (McCall and Simmons 1969). The words *participant* and *observation* in the title caught my interest. I decided to take a few minutes to skim through the work. It is a collection of papers written about a major method of field research in social science. My interest in the subject matter quickly intensified as I begin the first paragraph of chapter one. "Surely no social scientific method has given rise to more criticism and controversy in the past twenty years than has the broad approach known as participant observation."

These are pretty strong words for me. I was now hungry for both a definition of *participant observation* and an explanation of why the methodology would elicit such criticism and controversy.

Browsing further, I found that participant observation was a methodology of social science that employed a combination of methods and techniques: ". . . a style of research characteristically used for seeking analytic descriptions of complex social organizations. This style emphasizes direct observation, informant interviewing, document analysis, respondent interviewing, and direct participation and is made possible in large part by repeated, genuinely social interaction with the members of the organization under study" (Strauss et al. 1969, 26–27).

It is also a methodology in which: ". . . The observer participates in the daily life of the people under study, either openly in the role of researcher or covertly in some disguised role, observing things that happen, listening to what is said, and questioning people, over some length of time" (Becker and Geer 1969, 322).

With regard to the ". . . *more criticism and controversy in the past twenty years*," one has to consider the 1969 publication date of McCall and Simmons. In 2012, I found little evidence of participant observation studies in K-12 education in the existing literature and thus little to foster criticism or controversy—perhaps my work here will rekindle that fire.

In their book, *Designing Classroom Research: Themes, Issues and Struggles*, educational anthropologist Margaret Eisenhart and educational psychologist Hilda Borko state: "It is painfully obvious that much classroom research never finds its way into schools or classrooms, at least not as the researchers intended" (Eisenhart and Borko 1993, 75). The authors further state: "For classroom research to have an impact on educational practice, findings must be made accessible to practitioners in a timely fashion and in a form that can be understood and translated into action" (p. 76). To this end, the authors dedicate significant real estate to discussion of the difference between teachers and researchers and the various ways in which teacher-researcher collaboration can be improved. Although the term *participant observation* is not found in Eisenhart and Borko, the collaboration that the authors promote was yet another influence on my "prepared mind." It would be added to my readings of Jackson (1990); and, McCall and Simmons (1969).

Friday, February 4, 2011. As the school's data analyst, collecting, analyzing, presenting, distributing, and archiving data, as well as generating reports, are all functions of my job. Some of the most important sets of data that I attend to are our performance results on the state's assessments in mathematics, reading, writing and science.

On the preceding Monday, just four days previous, the Rhode Island Department of Education released the New England Common Assessment Program (NECAP)[1] scores for the Fall, 2010 administration of math, reading, and writing. Great news, we surpassed the state average in every category in all content areas assessed. We are hoping to make similar gains in the science assessment that would be administered in May—just a few months away. Unfortunately, the performance on the state's science assessment over the last three years could be best described by *embarrassing*. In 2008, only 5 percent of Beacon Charter High School's 11th-grade students taking the test, were identified as *proficient*[2]. We had reached a level of 14 percent in May of 2009 and this remained the same for May 2010. We were six percentage points below the state average of 20 percent. Science was now Beacon's weakest area where the NECAP assessments were concerned.

Still, we had learned a lot about preparing students for NECAP math, reading, and writing and had been quite successful in raising their test scores. It was now time to apply similar interventions in the area of science. I had made it a personal goal last year to try to do one project with every teacher on the staff. The "Water Wheel" project for Jen Wall's tenth-grade humanities classes was the first one I undertook—they were studying the Industrial Revolution; also, I had developed a GPA calculator program using EXCEL for Dana Pagliaro's ninth-grade Microsoft Office classes, and an online survey and analysis project for James MacDonald's eleventh-grade social studies classes as they were studying a state election. So, perhaps I could work with the science department to raise the NECAP scores under the aegis of that project.

Carolyn Taylor, our Director of Special Education, began working with Jessica Waters, the eleventh-grade science teacher, to integrate *Study Island*[3], a web-based science program, into the curriculum as one means of NECAP preparation. I had overheard that a training webinar had been planned, unbeknownst to me. For a person who likes to get involved in things, this was a bit disappointing. I signed up for the training webinar. In attendance were Carolyn Taylor, Jessica Waters, and Larissa Perkins (math teacher) and myself. Larissa was very familiar with Study Island as we had both used it last year to help raise the NECAP math scores.

After the session, we brainstormed how to best integrate Study Island into the eleventh-grade science curriculum. Given the relatively small size of the school, Jessica would be teaching all eleventh-grade science classes and all eleventh graders had to take chemistry. Jessica recorded all of our ideas into her laptop and thanked us for our input.

A silent "Aha!" moment! How about if I have the school register me as an eleventh- grade student, take chemistry everyday and do everything that the students are asked to do? I would attend every class and explore student engagement, teacher best practices, and hopefully support Jessica Waters in ways that could help boost our NECAP science scores. In other words,

I would perform ‘participant observation’ research. This would go beyond Philip Jackson’s tactic of observing from the back of the classroom and actually place me as a native in the village. Jackson’s goal was to deliberately remain neutral, to refrain from trying to answer such questions as: How can one reach those students who have all but given up on school? How should teachers be evaluated? How can we help today’s youngsters develop a love of science and mathematics? (Jackson 1990, xv). These were precisely the questions I hoped to explore since student engagement and teacher best practices were, for me, two areas of interest where long-term research was concerned. This was an opportunity for me to collaborate with a colleague with the aim of addressing some of the issues confronting educators today at both the local and national levels. These included: engaging students as a means of decreasing the number of drop outs; getting them prepared for the Rhode Island state science exam; increasing the scientific literacy of all of them—both those who ultimately pursue advanced degrees and careers in science-related fields, as well as those who pursue other disciplines; broaden the participation of women and minorities in science fields—a void that badly needed to be filled; and to prepare students for college so that they would not have to take remedial courses in their freshman year—a major problem in post-secondary education. I also sought to find an efficient means of using data to inform instruction; to provide a view of teaching and learning from the students’ perspective; and to identify teacher best practices that can be shared with other teachers. It was my hope that these would be of some value to pre-service teacher education. Finally, I wished to add to the research literature on the methodology of participant observation in K-12 education as that literature is still not extensively developed. And, it was my belief that my research on these questions could ultimately be generalizable to other content areas.

As math department chair, I was teaching only one class this semester, so I had the time. But could I get permission from administration to do this research? And, I really didn’t know Jessica Waters, the chemistry teacher that well. I had been on the search committee that interviewed and hired her a little over a year ago; however, I had never been in her classes. Would she agree to such an invasion? My own thinking was that, if I were in Jessica’s place, I probably wouldn’t, or at least I would participate in it only reluctantly. Eisenhart and Borko’s discussion of teachers’ and researchers’ differences came to mind, something they referred to as he “Great Divide” and their subsequent discussion of teacher-researcher collaboration (Eisenhart and Borko 1993, 79–81).

After the webinar, I ran up to the fourth floor to catch Dr. Pilkington, the school’s Principal at the time, to seek permission from him to do this. I found him in conversation with Mike Skeldon, the school’s Academic Dean and

Principal in Residence (i.e., Beacon's future Principal) and quickly explained my idea to both of them. They didn't exactly say "yes" or "no." Mike Skeldon said I wouldn't last 10 days. Dr. Pilkington gave me 30 days. So, not hearing a definite "no," I took their responses to mean that they approved of my adventure. I left and ran to Robin Murphy's office. Robin wears a couple of hats. She is both the Dean of Students and head of the guidance department. I explained my research to her and noted that I needed her to schedule me into Jessica Waters' Chemistry Block 1 class on *PowerSchool,*[4] the school's web-based school management system. She asked me to give her a day or so and she would schedule me in time for the start of the semester.

Next, came the difficult part, namely, trying to explain to Jessica that I was going to be conducting participant research in her class—that I would be signed in as a student and would perform all the tasks required of her students. She would have to treat me just like any other student in her class. I found her in her classroom and told her of my plans. I'm not sure what she was thinking, but like Dr. Pilkington and Mike Skeldon, I didn't hear a "no," so, to me, this meant "yes."

So, everything was set to go. I did, however, have one concern. My only blueprint on participant observation was from a book published 42 years ago. It was based on a research method that editors George J. McCall and J. L. Simmons had noted was highly criticized and controversial—even at that time. I therefore needed to investigate a more current view of this methodology. This I found in *Participant Observation: A Methodology for Human Studies* (Jorgensen 1989). Unlike Philip Jackson's pioneering work in his *Life in Classrooms*, in which he claimed to have begun his research with virtually no blueprint in hand, I had a few key resources: Jackson ([1968] 1990); McCall and Simmons (1969); Jorgensen (1989); and, Eisenhart and Borko (1993).

Of course, key to all of this was the willing participation of chemistry teacher Jessica M. Waters. I am extremely grateful to her for opening her classroom to me and for the opportunity to observe a master educator at work. I am especially pleased that my observations of engaging teaching and best practices during my 79 days in her Chemistry Block 1 class have since been validated by her receiving the Peter McWalters 2012 Professional Educator Award and being named the Rhode Island 2013 Teacher of the Year.

By way of reflection on this endeavor, Jessica and I believe that our collaborative work offers, at least in some part, valuable insights to all of the various stakeholders in education, particularly to those concerned with learning at the high school level. I here have in mind teachers of various content areas (not limited to the sciences), pre-service teachers, administrators, doctoral candidates in education and educational researchers.

Hopefully this background discussion will provide the reader with some insight into the motivations for this book. Enjoy the read! —J. D. B.

NOTES

1. In response to the requirements put in place by the federal NCLB (No Child Left Behind Act), Rhode Island partnered with Vermont and New Hampshire to develop standards and relate assessments to them. The NECAP consists of separate standardized assessments in the content areas of mathematics, reading, writing, and science and assesses student proficiency levels relative to these standards. At the high school level, eleventh graders take the mathematics, reading, and writing assessments in October with scores made public around February. They take the science assessment in May with the scores announced in September. The Common Core State Standards (CCSS) will replace the NECAP standards and assessments in mathematics and reading, with a transition at the high school level beginning in 2012. However, the NECAP Science standards and assessments will remain in place while the Next Generation Science Standards (NGSS) are being developed.

2. There are four levels of NECAP performance proficiency: *Substantially Below Proficient* (Level 1), *Partially Proficient* (Level 2), Proficient (Level 3), and, *Proficient with Distinction* (Level 4).

3. *Study Island* is the flagship product of Archipelago Learning, a PLATO® Learning company, and leading subscription-based software-as-a-service (SaaS) provider of supplemental education products. Founded in 2000, Study Island helps students in K-12 meet state-specific, grade-level academic standards in an enjoyable and engaging manner. It uses standards-based instructional, practice, assessment, and productivity tools and deliver them via web-based platforms.

4. *PowerSchool* is a widely used web-based student information system produced by Pearson Education, Inc. It supports 10 million students in all 50 states and in over 65 countries. PowerSchool enables educators to make timely decisions that impact student performance and, at the same time, to create a collaborative environment for parents, teachers and students. Beacon uses PowerSchool for managing data on centralized enrollment, attendance management, tracking and notification, assessment reporting, and other things.

Acknowledgments

Thanks to Carolyn Taylor (Director of Special Education) and Ann Brockmann (Special Education Teacher) to whom I would occasionally give drafts of chapters for review and feedback.

Special thanks are also due to the student members of the Beacon Charter High School Photography Club, Shawn Reynolds, Brad Cote, and Amanda Belisle for their images of various items of student work throughout this project; also, to Photography Club Advisor Ms. Suzanne Russell for her expertise and guidance.

My gratitude also goes to my editor, Deborah S. Nash, for her review and revision of the work and suggestions on substantive changes to it. Her efforts helped me clarify my thinking on a number of points and made it possible to move the project from manuscript to finished product with ease.

I offer my sincere thanks to my colleague, Jessica M. Waters, for allowing me to be a guest in her classroom for so many days. Her extraordinary cooperation and collaboration were key to the success of this project. I learned quite a lot about student engagement and teacher best practices in her Chemistry Block 1 class. And, what were the odds that the class I chose for the research would be the one being taught by the future Rhode Island 2013 Teacher of the Year? I cannot think of anyone more deserving of such an honor and I wish her the best in all her future endeavors.

Finally, Mrs. Waters and I would like to also thank our colleagues, the faculty and administrators at Beacon Charter High School, especially Dr. Robert Pilkington and Dr. Michael Skeldon for being supportive of our research.

Introduction

PERILOUS TRAJECTORIES TOWARD AN UNDEREDUCATED SOCIETY

Between 1.2 and 1.3 million public high school students, or nearly one-third of them, drop out of school each year. That is equivalent to one out of three students every 26 seconds 24/7, 365 days of the year. The average graduation rate for the fifty largest US cities is a discouraging 53 percent, the national average being 71percent. Urban areas are the most problematic as they contain our most vulnerable youth and communities—low-income minority, single parent children (Swanson 2009). Nationally, nearly one half of all black, Hispanic and Native American children fail to graduate with a regular high school diploma (Bridgeland, DiIulio, and Morison 2006; Swanson 2009). Moreover, the United States ranks eighteenth out of twenty-four industrialized countries worldwide in graduation rates.

The forecast for dropouts is dismal. As compared to their peers, they are more likely to be unemployed, live in poverty, receive public assistance, be incarcerated, be unhealthy, and, ultimately to become single parents with children who themselves drop out of school. In fact, one national study showed that dropouts comprise nearly half of all the heads of households on welfare. Another one revealed that 82 percent of America's prisoners are high school dropouts.

The economic consequences of leaving high school without a diploma are severe: on average, high school dropouts earn $9,200 less per year than high school graduates, and about $1 million less over a lifetime than do college graduates (Bridgeland, DiIulio, and Morison 2006).

Sadly, many drop out within two years of graduating high school and the majority of them have grades of C or better. Among the many reasons why students report they drop out: classes were not interesting to them, they missed too many days and could not catch up; they spent time with people who were not interested in school; they had too much freedom and not enough rules in their lives; and they were failing in school (Bridgeland, DiIulio and Morison 2006).

With regard to the measures of our schools' performance, the results of the 2009 NAEP (National Assessment of Educational Progress) also known as *The Nation's Report Card*, show that roughly 75 percent of US eighth graders are not proficient in mathematics when they complete eighth grade. Furthermore, 62 percent of US high school seniors fail to read at a proficient level and 26 percent of these fail to read at even the most basic level. In mathematics, nearly 75 percent of seniors fail to perform at a proficient level and 36 percent of these perform below the basic level. The NAEP results indicate that too many high school seniors are unprepared for college-level math and reading.

Such clear evidence of unpreparedness for college study makes it clear why an ever-increasing number of high school graduates have to take remedial math and English classes when they enter college as freshmen. They must pay for these courses, but do not earn credits toward their degree by completing them. Needless to say, we are on a perilous trajectory here. It is not unrelated to the teaching of grammar and the problems associated with it as these are clearly identified in college professor David Mulroy's 2003 work, *The War Against Grammar*. In the first chapter of this work, "America the Grammarless" he outlines his findings. One finding that has been disturbingly consistent is that many college English students are seriously deficient in knowledge of basic grammar. He describes having devoted entire semesters to trying to teach parts of speech, sentence diagramming, and the conjugation of English verbs to groups of college freshmen. At the end, the few students that could identify parts of speech were those who already had some degree of familiarity with them when they began the course. For the rest, professor Mulroy states, "It was like trying to teach table manners to a motorcycle gang" (Mulroy 2003, 4). "Am I alone in finding such ignorance remarkable?" he asks. In his view, however, students are not to blame because, at some point in the past, teachers decided to stop teaching grammar. There is, in my view, a clear parallel between many of Professor Mulroy's experiences and my own with teaching high school freshmen algebra students. They often enter my class two to three years behind in basic mathematical skills—not knowing, e.g., multiplication tables, common decimal equivalents of fractions and percentages, measurement conversions, etc. Several of them count on their fingers or stare at the ceiling when asked to subtract nine from seventeen or divide sixteen by four! Like work on parts of speech, sentence diagramming and conjugation of English verbs, somewhere in those early grades the fundamental building blocks of mathematical problem solving have been thoroughly de-emphasized, ignored altogether or supplanted by other course content.

It is now a well-known fact that, on international assessments, US students lag behind those of the highest performing nations. The results for the 2009 PISA (Programme for International Assessment) show that, of the thirty-four countries of the OECD (Organization for Economic Co-Operation and Development), US fifteen year olds ranked twenty-fifth in mathematics, fourteenth in reading and seventeenth in science (OECD 2010).

"Employers in many industries lament that job applicants lack the needed mathematics, computer, and problem-solving skills to succeed, and international students fill an increasing portion of elite STEM [science, technology, engineering, mathematics] positions in the United States. Indeed, in 2007, 'international students constituted more than a third of the students in US science and engineering graduate schools,' and more than 70 percent of those students currently remain in the United States after earning their degrees" (NRC 2011).

The teaching profession has changed drastically in recent years. Reasons include more rigorous teacher evaluations; teacher employment decisions (e.g., certification, retention, merit pay) linked to their students' performance on state assessments; more testing days making for less instructional days; more training on new technology; new standards (e.g., Common Core State Standards, Next Generation Science Standards); increasing encroachment into vacations; and in some cases longer work days. "Forty percent of K–12 teachers are currently 'disheartened' with their job . . ." (Hargreaves and Fullan 2012, 5).

All of these are very sobering statistics, all the more so when one considers the intensive effort that has ostensibly been put toward educational reform over the last several decades. We may be in the same predicament as Alice in Wonderland—at a fork in the road! And, the pathways available to us have very different consequences indeed!

HELPING ALICE CHOOSE A PATH

In the 1989 film, *Field of Dreams* (nominated for three Academy awards), an Iowa farmer heard a voice from his cornfields: "If you build it, they will come." He interpreted this as a command to build a baseball diamond there. Subsequently, he plowed over his fields and built a ballpark, then deceased members of the Chicago Black Sox show up. In education, we may be able to build the finest structures, but a school has to be more than brick and mortar, or, the students will not come.

In December 2009, I successfully defended my doctoral dissertation: *Exploring Some Inattended Affective Factors in Performing Nonroutine Mathematical Tasks*. One of my major findings was the significance of something I refer to as *math-aesthetics* as these concern student engagement in problem solving. They were more apt to be engaged in performing a task if they were attracted by some aesthetic feature of it—perhaps an interesting picture, their own name used in a word problem, or a hands-on activity such as origami. Further, my research showed that socio-economic status, gender, and ethnicity were not factors that were relative to math-aesthetics. Given that "being bored" is the number one reason that students cite for dropping out and that many of them were, in fact, passing their classes when they dropped out (Bridgeland, DiIulio, and Morison 2006), the opportunity to research student engagement has become a focus of mine.

In their 2012 book *Professional Capital: Transforming Teaching in Every School*, authors Andy Hargreaves and Michael Fullan focus on the question of getting and keeping good teachers and ensuring quality teaching. They describe two current schools of thought on teaching—one driven by a view of *business capital*, the other by *professional capital*.

From a business capital perspective, education is seen as ". . . a big new market for investment in technology, curriculum and testing materials, and schools themselves as for-profit enterprises. In the estimates of some multinational moguls, this is a massive $500 billion market" (p. 2). The authors describe the business capital view of teaching as acknowledging that good teaching may be emotionally demanding, but is technically simple; good teaching is quick study requiring only moderate intellectual ability; good teaching should be driven by high performance data about what works and where best to target one's efforts; good teaching comes down to enthusiasm, hard work, raw talent, and measurable results; and, good teaching is often replaceable by online instruction (p. 14). The authors are highly critical of alternate certification programs that ". . . cram the preparation of teachers into just a few short weeks and see many of their qualified teachers leaving after only a few years in the job" (p. 6).

In contrast, the authors describe and promote the professional capital view that good teaching is perfected through continuous improvement; good teaching involves wise judgment informed by evidence and experience; good teaching is a collective accomplishment and responsibility; and good teaching maximizes, mediates, and moderates online instruction (p. 14).

Hargreaves and Fullan state that the professional capital approach is a strategy that has already been adopted by the highest performing economies and educational systems in the world. They argue that professional capital should be thought of as the product of three other types of capital—human, social, and decisional. Human capital pertains to the "talent" of individuals and is based on the belief that they have the power to change the system; social capital refers to the idea that interactions among teachers and between teachers and administrators, i.e., groups, teams, and communities, are far more powerful than those among individuals. In their view, therefore, collaborative efforts ". . . focused on student learning make a "large and measureable difference in student achievement and sustained improvement" (p. 3). Improving social capital increases human capital, but the reverse is not true. The third type, decisional capital, refers to the pros who have the ability make decisions in complex situations. Hargreaves and Fullan (2012, pp. 22–23) conclude that "Teaching like a pro means":

- Continuously inquiring into and improving one's own teaching.
- Planning teaching, improving teaching, and often doing teaching not as an individual but as part of a high-performing team. It means developing shared professional capital within an organization and community.
- Being part and parcel of the wider teaching profession and contributing to its development. To grow, professional capital must circulate freely, energetically, and openly. This means rethinking how teachers work with, support, and also challenge their colleagues.

The collaboration between Mrs. Waters and myself described in the following chapters is highly supportive of the view that social capital can improve human capital. We both believe that through our shared ideas, support, and feedback, in addition to acquiring increased social capital, we have grown individually where teaching skills are concerned. We have gained significant insight to student engagement and teacher best practices and these are findings that we believe can positively contribute, in at least some small way, toward a correction of the current course of our perilous educational trajectories—high dropout rates, student underperformance, and unpreparedness for college and career. This book represents our attempt to be part of a wider teaching profession and contributing to its development.

SETTING THE STAGE

> Rather than denying personal interests and values, the methodology of participant observation requires an awareness of how these thoughts and feelings influence research. By reporting personal interests and values, other people are able to evaluate further the influence of values on your findings.
>
> —*Danny L. Jorgensen, Participant Observation: A Methodology for Human Studies*

It is with this in mind that I provide vignettes of my personal experiences throughout this book—events that occurred prior to my participant observation research with Mrs. Waters.

I spent 79 consecutive days in her Chemistry Block 1 class. My time there spanned the complete semester-long course from Day 1 through the concluding Day 79. During this time, I filled two journals with field notes—nearly 400 pages in total—recordings and interpretations in real-time. This book represents the fruits of my research. Its style is not too much unlike Philip Jackson's *Life in Classrooms.* He described his work as a mélange wherein "Descriptions of empirical studies are interlaced with speculative asides; tabular materials sometimes share the page with the most unquantifiable assertions" (Jackson 1990, p. xxi). I too have interlaced quantitative and qualitative perspectives. From the wealth of data collected, I was able to derive major themes. These I deal with as separate chapters. Here is a brief overview of them:

In chapter 1, "The Natives and the Village," I outline school demographics and a detailed look at the classroom environment. Chapter 2, "Going Native: My First Ten Days in Chemistry," provides a detailed look at my first ten days in Mrs. Waters' class. Research has shown that very often the tone for the entire school year is set within the first two weeks in the classroom. Chapter 3, "The Aesthetics of Learning," explores student engagement and chapter 4, "Mrs. Waters' Formal Lab Report," discusses this assignment. It illustrates the point that the use of data can drive instruction, improve lessons, and subsequently raise student achievement. Chapter 5, "Mrs. Waters' Exit Survey: What does the research say?" presents the results of an exit survey which sought student feedback on the question of their perceived engagement with the various tasks she routinely assigned during the semester. Chapter 6, "The State Science Exam," discusses the efforts made by way of preparation for the state's science assessments and concludes with a summary of the results of Mrs. Waters' efforts. Chapter 7, "A Taxonomy of Time," focuses on the precious commodity of classroom time and includes a detailed list of reasons why actual learning time is decreased. In Chapter 8, "Mrs. Waters' Best Classroom Practices," I present a comprehensive list of teacher *best practices*—those observed in her class as well as those that research as brought to light. Chapter 9, "Teacher of the Year—It's the Climb," presents a rare look at a teacher's journey from being selected by her peers for the district's "Teacher of the Year;" to becoming one of three state finalists in this competition, to being named the Rhode Island 2013 Teacher of the Year. This is an unprecedented inside look in real-time at the pressures of such a climb. It is also an amazing story. These events occurred after my research in her class had begun. Her being named for this award came as a complete surprise to me. I just happened to be in the right place at the right time since I was able to record the entire process as it unfolded. This chapter also contains a discussion of Mrs. Waters receiving the Peter McWalters Professional Educator Award. In chapter 10, "Reflections on the Research," I have addressed traditional concerns about the validity of ethnographically based research by seeking feedback from student participants as well as from Mrs. Waters on their perceptions of my presence in the culture of their classroom. Their reflective responses, stated in their own words, are also included. In the Epilogue, "What We Have Learned and What Next?" I offer my own reflections on my research and summarize my thoughts on this.

Finally, the appendix "A Prime Connection?" describes an "Aha!" moment I had while I was doing one of Mrs. Waters' *flexible* homework assignments. I had chosen the topic of "electron configurations" and saw a pattern of prime numbers—numbers that have, as it turns out, been a matter of curiosity for two thousand years.

ADDITIONAL COMMENTS FOR PURPOSES OF CLARIFICATION

First, from this point forward and throughout this book, I refer to my collaborator on this project and colleague, Jessica M. Waters, as Mrs. Waters because two students in her chemistry class are also named Jessica. Where applicable, I distinguish between these two students by using Jessica P. and Jessica G. In addition, if I am truly a student in her class, I should be addressing her as Mrs. Waters out of respect for her as my teacher.

Secondly, when I began this project in the Spring of 2011, the Principal of Beacon Charter High School was Dr. Robert Pilkington. Michael Skeldon was the Academic Dean and in the process of completing the Principal in Residence Network (PRN)[1] program. In the Fall of 2011, Dr. Pilkington, who is considered by many to be a pioneer of the charter school movement in Rhode Island and founder and past President of the Rhode Island League of Charter Schools, left Beacon to assume command of his newest venture, the Rhode Island Nurses Institute Middle College. This is the first nursing middle college in the country. Consequently, Michael Skeldon became Principal of Beacon for the 2011–2012 school year. He completed and was awarded his Doctorate in Education (EdD) from Johnson & Wales University in May, 2012—just before this book was completed. Thus, I refer to Dr. Skeldon as Mike Skeldon, or Mike throughout, not out of lack of respect, but because he has been a close friend since I joined Beacon in 2005.

A final note before class begins: Throughout this book, the students' actual first names are used (permission was granted). In no case are their last names included. Of the seventeen members of Mrs. Waters' Chemistry Block 1 class, to which this book is dedicated, only one student chose not to have her first name mentioned. I have honored her request and used the pseudonym "Ruby" instead. Further, in cases where a participant, or any individual mentioned, could possibly construe as embarrassing an event, which is being described in this work, I have replaced real names with pseudonyms.

NOTE

1. Principal Residency Network (PRN) is a non-traditional, state approved, research-based leadership preparation program. Its aim is to train participants for principalships statewide. Aspiring leaders serve a one-year internship with an outstanding mentor principal and are awarded principal certification upon successful completion of program requirements.

Chapter One

The Natives and the Village

> We begin to see, therefore, the importance of selecting our environment with the greatest of care, because environment is the mental feeding ground out of which the food that goes into our minds is extracted.
>
> —*Napoleon Hill, The Law of Success, Volume III: The Principles of Self-Creation*

"IT STINKS, IT LEAKS AND THE WINDOWS FALL OUT"

"It stinks, it leaks and the windows fall out" was the title of a lengthy article by staff writer Gina Macris. It appeared in *The Providence Journal* (Metro Edition), Rhode Island's leading newspaper, on Friday, June 6, 2003—just seventeen days short of the end of my second year of teaching. Sadly, the reference was to the building in which I was teaching and the article included a five-by-eight-inch picture of my third floor classroom. But, alas, the title did not refer to a smelly teacher or smelly kids! Rather, the subject was the abhorrent conditions of the Fogarty Building located at 111 Fountain St. in downtown Providence. It was used as temporary housing for three small public high schools: The Providence Place Academy (PPA); The Providence Academy of International Studies (PAIS); and, the Health, Science, and Technology Academy (HSTA). At the same time, a brand new complex was being built off of Thurbers Avenue, a few miles away on the South Side.

In chapter 8 of this work, "Mrs. Waters' Best Classroom Practices," I discuss my first few days of teaching in this building. Briefly stated, I was the only math teacher for the Providence Place Academy—teaching ninth and tenth graders at the Fogarty Building every other day and eleventh and twelfth graders at the Providence Place Mall on the other days. This public high school focused on a marketing and retail curriculum with its eleventh and twelfth graders serving internships at various stores located in the mall—an innovative concept and the first of its kind in the nation. There were three modern classrooms located below street level at the mall. This limitation on the number of classrooms at the mall site had necessitated utilization of the condemned Fogarty Building, just a short walk for our ninth and tenth graders. It was to be razed and replaced by a two hundred and fifty-suite, thirteen-story hotel—a $57 million project that had been put on hold.

"It stinks" referred to the smells emanating from the mold and musty conditions caused by the leaky roof and walls. The school had a history of frequent teacher sick leave because of the sinus and respiratory illnesses caused by these conditions. Students had vandalized the bathrooms and torn hand towel and toilet paper dispensers from the walls. Even doors had been ripped off the bathroom stalls. Sometimes weeks went by without any hand towels or toilet paper in the bathrooms. At least once a week, students plugged up the toilets and urinals to such a degree that a lake formed in the school's wide hallway. "Attention shoppers! Spillage in aisle 3." There is nothing like the feeling of that wetness that seeps into the soles of your shoes! It oozes through your socks and the moisture even finds its way between your toes! By the way, faculty and students shared the same bathrooms. The problem of standing water was so bad that some students had chosen to make their own bathrooms on the vacant second floor. Piles of human feces in several areas and handprints on adjacent walls were evidence of this. At the bottom of page C3 of the *Journal* article, there appeared a rather gross picture of the girl's bathroom with the caption: "TRASHED: The girls' bathroom on the third floor of the school has one small sink, no paper towels and no hot water." The picture showed walls that were completely covered with graffiti. In fact, my first *professional development* (PD) experience in education was with the entire faculty from all three Fogarty high schools in a meeting facilitated by local detectives. They had photographed artwork in the bathrooms and educated us on how to recognize gang signs and interpret their messages.

The article's phrase "and the windows fall out" was a reference to several incidents in which large classroom windows had indeed fallen out of their casings and landed on the street below. It had correctly reported that one of them fell inward and hit a student. My third-floor classroom was missing a window for two consecutive weeks in the middle of a cold New England January.

During that first year, complaints had been made to the school department about this, but no response was forthcoming. I often wondered if corrective action would have been accelerated had our union building representative been required to teach in the Fogarty Building rather than in her plush classroom over at the mall site. It was rumored that she had the word *grievance* tattooed on her left buttock! But considering that I was the new kid on the block, a first year teacher, and working on an *emergency teaching certificate*, I did not have much leverage in arguing for change.

These conditions persisted well into my second year at Fogarty. However, at that point, a few hundred students from the PAIS and HSTA high schools were now sharing the building. Hearing that they might still be in this building the following school year because the new school would perhaps not be finished, several students from PAIS photographed the conditions and submitted them to *The Providence Journal*—hence, *Journal* staff writer Gina Macris' article. Just prior to the June 6, 2003 publication, a report of an investigation by Rhode Island Department of Labor and Training into conditions at the Fogarty Building had noted the hazardous conditions—the falling windows and the mustiness and mold caused by the leaky roof. It had cited a total of thirteen health and safety violations. Problems included exposed electrical wiring, missing fire extinguishers, non-functional emergency lighting, and inadequate food preparation and serving areas.

The picture of my classroom mentioned earlier showed the large, shredded vinyl accordion shaped retractable curtain that separated it from the English classroom. At some point prior to my hire, students had ripped holes in the curtain.

Monday, June 9, 2003, 7:00 a.m. I arrive at the Fogarty Building only to find all of its entrances surrounded by yellow crime tape and several Providence police vehicles strategically positioned outside. Yes, on the very next school day after the newspaper article was published, the Providence school department's Superintendent, Melody A. Johnson, along with Mayor David N. Ciciline had declared the building unsafe and unacceptable under current conditions. It was therefore not allowed to operate as a school. Subsequently, we were all herded a short distance up the street to the Dunkin' Donuts Center (also referred to as the "Dunk," formerly the Providence Civic Center) where temporary canvas-partitioned classrooms had been quickly constructed over the weekend to accommodate the approximately 380 displaced students, faculty, and staff. This would be our campus for the remaining two-plus weeks of the school year.

Fortunately, I had a great relationship with my principal, Don Pastine. He had some sort of connections with persons of importance at the Dunk (Maybe it was an Italian thing!) and so was able to get me a great classroom. It was the Providence College Basketball team (*The Friars)* locker room and was outfitted with a wall-mounted whiteboard with a full-court layout stenciled on it for play strategizing. The name PROVIDENCE COLLEGE was painted across the top. I still have a photograph of an algebra class that I taught in the locker room. It clearly shows me at the whiteboard and one can see the backs of the students seated in front (highly engaged as I remember), and two urinals on the right side of the whiteboard.

The remaining few weeks of that school year were dedicated to preparing and administering the final exams. We never returned to Fogarty.

Late that summer, I received a schedule and notice to report to Central High School across town—the mall school was being restructured and integrated into Central. However, I had already accepted a teaching position in another Rhode Island school district. PAIS and HSTA would eventually move into the new building on Thurbers Avenue.

. . . EIGHT YEARS LATER: THE VILLAGE

I was born and raised in Rhode Island, the nation's smallest state. It is only about forty-eight miles long and thirty-seven miles wide and is bordered by Massachusetts on the north and east, Connecticut on the west, and the Atlantic Ocean on the south. Known as *The Ocean State*, Rhode Island boasts over 400 miles of coastline. Beacon Charter High School for the Arts is located in Woonsocket, just a few miles south of its northern border with Massachusetts.

I've been making the 24.88-mile trek up I-295 (western bypass around Providence) to Beacon every day for five-and-a-half years now. Despite the small size of the state, I had never been to Woonsocket prior to joining the faculty in the Fall of 2005.

During a hundred year period from the early 1800's to early 1900's, the city was one of the largest textile manufacturing centers in the United States. Cotton fabric was a major export. I am reminded every day of this history as I pass by the many factories and mills, long since abandoned, along the rivers that once powered textile machinery. I also pass many old churches with great architecture—the large number of such intricate structures being indicative of a once densely populated and vibrant community. Sadly, Woonsocket has not fared well economically in recent years. Just about every morning, within a thousand yards of Beacon's front door, I drive past a group of homeless individuals who are just leaving shelters after their brief nightly escape from sleeping in the open air. And, as I am typing these pages, the local Woonsocket public school department is $10 million in debt and the city is contemplating declaring bankruptcy.

Beacon Charter High School for the Arts is located in a four-story historic building (see figure 1.1) at 320 Main Street in downtown Woonsocket. It relocated to Main Street from its former George Street location in October, 2006. Formerly occupied by the ARC of Northern Rhode Island (The Home-

Figure 1.1. Beacon Charter High School for the Arts 320 Main Street; Woonsocket, Rhode Island.

stead Group), the building's second and third floors had to be refitted from an open-floor factory layout in order to facilitate the creation of classrooms. It was purchased by Beacon in Summer 2011 and is located diagonally across the street from the historic Stadium Theatre.

Beacon, which opened in 2003, is a charter school (grades 9–12) whose courses of study combine a college preparatory program with an arts program that focuses on three disciplines: culinary arts, theater arts, and visual arts. The founders of the school envisioned a unique high school experience that would prepare student-artists for a career in art related fields. The charter school model offered such an opportunity.

Charter schools are public schools authorized by the State of Rhode Island to operate independently of many of the rules and regulations imposed by the state and local district. In exchange for this greater autonomy, charter schools assume greater accountability for meeting student achievement goals. Their charters can be revoked for inadequate performance. However, they provide greater freedom of choice to parents and students seeking alternatives to what traditional public education has to offer. Because of their autonomy, charter schools are encouraged to be innovative and creative in their educational philosophies. They are also expected to be laboratories and to represent the vanguard where improving and expanding opportunities within public education (Office of the Commissioner 2012) are concerned.

Like all public schools, charter schools make education available to anyone free of charge. In Rhode Island, as in many other states, there are more applicants than seats available. By law, a lottery must be held to fill the available seats. For each of the last several years, Beacon has had a waiting list of students hoping to enroll.

SOME DEMOGRAPHICS

"The methodology of participant observation requires that the researcher become directly involved as a participant in peoples' daily lives" (Jorgensen 1989, 20).

Although the school's focus on art is what attracts students, a significant number of Beacon students and parents choose it because it is regarded as a safe and caring environment. The school draws its two hundred and twenty-five students from school districts all around the state. However, the majority of them typically come from the closest districts, largely from Woonsocket, in part because of transportation issues. And, they come with diverse experiences and abilities.

I chose Mrs. Waters' Chemistry Block 1 class for this research partly because I was available each day for this class, and partly because, of her three classes, Block 1 contained those students who were the highest performing academically based on their math and reading scores. If I were going to study student engagement, the natives in Block 1 would be the most likely to be engaged.

At the start of the semester, seventeen students (excluding myself) registered to take chemistry. Of these, sixteen were female and one male. Four of them were studying culinary arts, five theatre arts, and eight visual arts. Only one was a *special needs* student with an IEP (Individualized Education Plan)[1]. It is common for educational researchers to use the number of students who qualify for *free* or *reduced paid* lunches as an indicator of socioeconomic status of the school's population. In this cohort, of the seventeen students, 47% (8/17) qualify, an indication of a high rate of poverty in the school. With regard to ethnicity, fourteen students self-report that they are white, two, native Hawaiian/other Pacific Islanders, and one, African American. The average age of these eleventh-grade students is sixteen years and eleven months (as of February 4, 2011—the first day of class). The seventeen students represent seven different school districts in Rhode Island (seven students are from Woonsocket).

MRS. WATERS

Mrs. Waters graduated from Rhode Island College in 2008 earning a Bachelor of Arts Degree in Secondary Education with a biology major and chemistry minor. She joined our faculty in the 2008–2009 school year and is currently in her third year of teaching at Beacon. She has taught physics, chemistry, human anatomy and physiology as well as SAT Prep.

She also co-founded and serves as advisor for our SADD (Students Against Destructive Decisions) chapter. SADD's mission is: "To provide students with the best prevention tools possible to deal with the issues of underage drinking, other drug use, risky and impaired driving, and other destructive decisions" (SADD 2012). She also has an advisory. Teachers at Beacon are assigned to a small group of students and serve as mentors to them throughout their high school experience. Mrs. Waters advisory has won Beacon's Annual Thanksgiving Food Drive three years in a row—further testimony to her dedication to positively shaping and influencing children's lives.

MRS. WATERS' TYPICAL WEEKDAY

Mrs. Waters (Jessica), age 33, is happily married to husband Chris and the mother of two daughters,[2] ten-year old Mackenzie and twenty-month old Madeline. Like many working mothers, she must balance the responsibilities associated with family and career. The following is a synopsis of her typical weekday as wife, mother, and teacher.

4:30 a.m. Mrs. Waters wakes up but admits to typically lying in bed for about fifteen minutes. She never uses an alarm clock.

4:45 a.m. She showers, does her makeup, and gets dressed as she has to get herself ready before she wakes the kids. She awakens and feeds the baby, then makes lunches for everyone. Next, she prepares breakfast—something as simple as cereal and a bagel, or perhaps something more complicated like eggs and cheese. Chris will later make coffee.

5:30 a.m. She wakes up her husband Chris and daughter Mackenzie. He needs to leave by 6:45 in order to catch the 7:07 train to Boston.

6:40 a.m. She leaves the house with both daughters. She drops daughter Mackenzie off at the carpoolers. Next she picks up her younger sister

Brittany who is a ninth-grade student at Beacon and, coincidently, a student in my algebra class this year. Mrs. Waters then drives to Smithfield and drops off baby Madeline at the daycare center.

7:05 a.m. She leaves the daycare center with sister Brittany and drives to Beacon Charter High School.

7:30 a.m. She arrives at Beacon with sister Brittany.

8:00 a.m.–2:30 p.m. She will teach three, eighty-minute long, chemistry classes, one of which is a low performing class and often requires multiple levels and types of preparation.

3:15 p.m.–4:30 p.m. She leaves school. Her departure time varies from day to day depending on her obligations to after-school meetings. She picks up Mackenzie in Cumberland, then drives to Smithfield to pickup Madeline from the daycare center.

4:30 p.m. She drops off Mackenzie at swimming (four nights a week) in North Providence. Mackenzie is often late and gets in trouble. She must be responsible for making sure she has her swimming gear in her mother's car each day because there will be no time to go back home to pick it up.

4:45 p.m. She goes home to make dinner for the family. However, only Madeline will eat at this time. Husband Chris won't be home until around 8:00 and will have dinner then. She dresses Madeline in her pajamas and ready for bed, but needs to pickup Mackenzie at swimming.

5:45 p.m.–6:00 p.m. She picks up Mackenzie and returns home.

6:15 p.m.–6:30 p.m. She rocks baby Madeline while reading to her to get her to go to sleep.

6:45 p.m. She now starts her schoolwork including correcting papers and preparing her lessons for the next day. This can take anywhere from two to three hours.

8:00 p.m.–8:30 p.m. Chris arrives home. Dinner is waiting for him.

10:00 p.m.–10:30 p.m. Mrs. Waters goes to bed.

Within this day, she must also find time to help Mackenzie with her homework and do household chores. There is no time for grocery shopping during the week—this has to happen on Sunday morning. She says that she seldom watches television and rarely goes out to eat or to a movie because of the complexity of her schedule during the week.

INTERVIEW WITH HUSBAND CHRIS

Technically, Mrs. Waters' husband, Chris, isn't a native (i.e., not routinely in the physical school building). However, given the long hours of Mrs. Waters' typical weekday, I couldn't help to wonder what he thought of the demands of the teaching profession. I knew he was the Chief Accounting Officer for a major firm working in global asset management, but I had only met him once—at the faculty picnic that Mike Skeldon (Academic Dean and future Principal) hosts at his house at the end of each school year. I couldn't pass up the opportunity to pose a few questions to Chris. He responded as follows:

> Q: What are your perceptions of the teaching profession?
>
> A: After closely observing my wife's career, I can say that good teaching doesn't just happen. Classroom excellence takes patience and dedication. I believe the quality of the teacher is the single most important variable influencing the education the student receives.
>
> Q: Are teachers paid fairly for the number of hours worked?
>
> A: As discussed above, good teaching is hard work. Again, just observing my wife, quality teachers spend a significant amount of time working outside of the classroom to stay prepared and organized. And, from my private-sector experience, it is fair to conclude that good teachers are not paid adequately for the number of hours they work. I feel the system would employ several more great teachers under the typical "pay for performance" compensation model.
>
> Q: Are there any negatives here where family life is concerned with being married to a teacher? How has it affected your job (if at all)?
>
> A: I can't think of anything meaningful. Jessica works very hard, and often very late, preparing lesson plans and gaining expertise on subject matter while caring for our three beautiful daughters[2]. Honestly, I believe she would be just as determined to achieve excellence in any profession. I am sometimes envious of her dedication to the students (and her summer schedule!).

MRS. WATERS' CLASSROOM

> Observation begins the moment the participant observer makes contact with a potential field setting . . . It is extremely important that you record these observations as immediately as possible and with the greatest possible detail because never again will you experience the setting as so utterly unfamiliar.
>
> —Danny L. Jorgensen, *Participant Observation: A Methodology for Human Studies*

With only 225 students enrolled at Beacon, two science teachers are enough to handle the required number of classes across all four grades. When science teacher Shayna Fox became the school's full-time reading specialist midway into the 2010–2011 school year, a long-term substitute science teacher had to be hired to complete the school year. This made Mrs. Waters the lead science teacher at Beacon. She inherited Shayna's classroom, the larger of the two science rooms located at the end of the second-floor hallway on the southeast corner of the building. A general layout of the room, as I first saw it on Day 1, is shown in figure 1.2.

The only entrance to Room 205, Mrs. Waters' classroom, is located in the northwest corner. My overall impressions, as I enter and take a cursory view of it, is that it is a warm, inviting place for learning science—far from the cold, uninviting classrooms with pane-less windows in which I was asked to teach in the Fogarty Building. With Jorgensen's quote in mind ". . . It is extremely important that you record these observations as immediately as possible and with the greatest possible detail . . .," I therefore take out the journal for my field notes and jot down the following:

Room 205 overlooks the intersection of Main and Church Streets. Both the south and east walls are banked with large windows. The south wall windows overlook Church Street and the teacher's parking lot; the east wall windows along Main Street provide a slightly angled view of the historic Stadium Theatre. It is diagonally across the street. Unfortunately, this view inherently includes the building directly across the street adjacent to the theatre that serves as a halfway house for recovering alcohol and drug users, perverts, and similar unsavory characters.

The large classroom windows extend upward to the ten-foot four-inch high ceilings and suck the sunlight and its heat from the sky to such an extent that temporary, ugly dark, partially opaque, sagging *drape "wannabes"* have been hung to mitigate against this effect. Without this makeshift window shading, any writing or projected media on the whiteboard, mounted on the west wall (considered the front of the classroom) becomes invisible.

Mrs. Waters has nine six-foot long, folding tables with laminated tops arranged in a U-shape. These provide space for two students per table. The opening of the U faces a sixteen-foot long whiteboard on the west wall. This configuration allows every student to have a clear view of the board. Also, she has an unobstructed view of each student; and, from within the U, a physical access to each of them since they are seated along its outside perimeter.

Her desk is arranged in an L-shaped configuration located in the northeast corner of the room. Two file cabinets abut the side of the desk. The top of the desk is clean and well organized. A tall grey metal cabinet, noticeably dented, is tucked into the corner behind her desk. It is stuffed with chemistry related glassware comprised mostly of test tubes and beakers.

Facing the west wall, there are two fish tanks on a table just to the right of the whiteboard with a poster entitled *Anatomy of an Atom* hanging over them. A large Periodic Table poster hangs above the whiteboard. A life-sized, one-legged skeleton stands to the left of the whiteboard just in front of a black metal bookcase in the southwest corner of the room. I see the missing leg, along with miscellaneous lab equipment, on a table nearby. For some strange reason, the black metal bookcase, originally designed to sit on the floor, has been haphazardly raised and placed on top of a wooden table. It is stacked full of chemistry textbooks. Next to the skeleton is a cheap Meade telescope (model #223S) on a tripod stand. It seems to be missing some parts. Two wooden bookcases, one tall, and one short, are located on the east wall (the rear of the classroom) overlooking Main Street. Both have seen better days. They are over-filled with an assortment of books, which causes the shelves to bulge. The books are tumbling out the back of the tall bookcase because its cheap, thin fiberboard backing has come unstapled.

An odd-shaped, double-decker computer table (see "S" in figure 1.2), overloaded with miscellaneous lab equipment (and one skeleton leg), is located along and against the south wall. This forms a very narrow corridor for students to walk behind the backs of other students seated on that side of the U.

There is also another obstruction to the flow of traffic around the room, namely, a large, oval shaped table positioned between the fish tank table and the classroom tables in the U (see "M," figure 1.2).

If one looks to the left when entering the room, there is a small jut in the wall, just large enough for a bulletin board to be hung.

At one point at the beginning of the school year, Mike Skeldon, Academic Dean and Principal in Residence, had suggested that Mrs. Waters 'make the room her own' because when he entered her classroom, he felt she was renting space. He was referring to the student artwork that had been painted on the walls during the period of Shayna's tenure there. She had tapped into her students' artistic talents to decorate the room the previous year. I continue to add notes and diagrams on this in my journal:

Below the whiteboard is an underwater scene including a fish, crab, squid, turtle, jellyfish, and porpoises. To the right of the whiteboard is a painting of the bow of a large ship, perhaps the Titanic, with a mermaid jumping out of the water. Above the whiteboard is a scene of outer space. It shows our sun and eight planets—poor Pluto is not included since it was downgraded to the status of a *dwarf planet* in 2006 and again in 2009. NASA had temporarily classified Pluto as a FLR (fairly large rock) until a better name could be worked out. Also, appearing at the bottom of the space scene is a tall, alien figure with a football-shaped head. Beyond the third rock from the sun is a car in flight shooting out flames. My guess is that it is Dr. Emmett Lathrop "Doc" Brown's *Back-to-the-Future* time traveling DeLorean from the 1985 award winning American science-fiction adventure film.

An elephant peeking through some African foliage is painted on the south wall between two large windows. Other wall paintings include: a banana

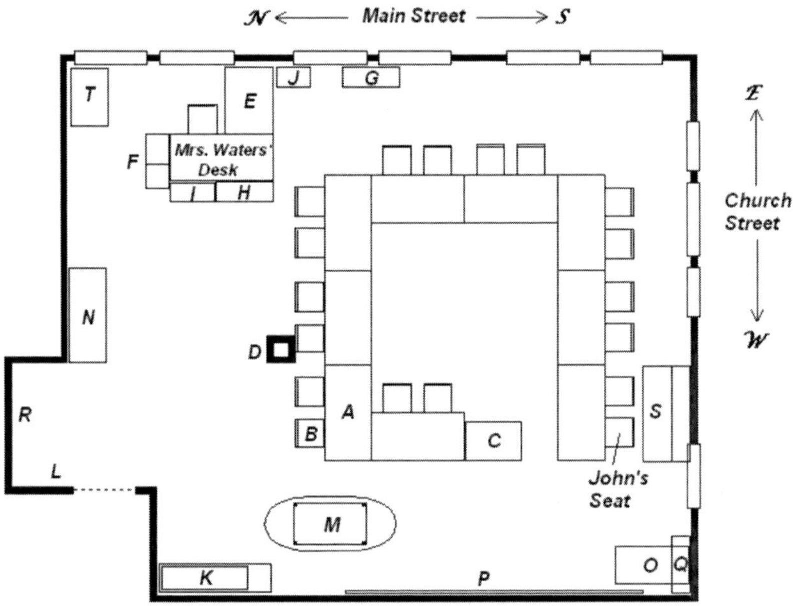

A. Classroom table: 60" x 30"
B. Chair
C. Projector table w/projector
D. Column
E. Mrs. Waters' desk
F. 2-Draw filing cabinets
G. Tall wooden bookcase
H. Metal bookcase
I. Short wooden bookcase
J. Chrome metal shelf unit
K. Fish tanks and table
L. Phone jack
M. Large oval shaped table
N. Computer table w/two computers and one printer
O. Table: 48" x 24"
P. Standard whiteboard
Q. Black metal bookcase for textbooks
R. Bulletin board
S Computer table w/miscellaneous lab equipment
T. Tall metal cabinet

Figure 1.2. Mrs. Waters' classroom (Day 1).

tree, with a monkey, in the southeast corner; a moose standing on the shore of a Canadian lake on the east wall between two large windows; a dinosaur on the north wall and yet another just around the corner of the wall near the bulletin board.

There is one column in the room near the left side of the U (see "D" in fig. 1.2). Its unfortunate position limits the size of the student's body that can comfortably fit into the seat near to it. Three sides of the column are painted with frogs and the fourth displays a colorful parrot. Plastic magazine holders stuffed with Scholastic *Science World* magazines are mounted on each of the four sides of the column.

Entering Room 205 and looking to the left and up toward the ceiling along the north wall, I see the partially finished faces of two famous scientists, both gazing downward. One is recognizable as Einstein, the other less discernible. Maybe it is Democritus, the Greek philosopher who was credited with developing the idea of the atom; or Charles Darwin, the famous English naturalist best known for his epic work *On the Origin of Species*. Above the door, there is an unfinished painting of a lion.

The Rhode Island state science standards are divided into three categories: Physical Science, Life Science, and Earth and Space Science. As I look around the room at the artwork, there appears to have been some strategy for the development of the artwork that was related to all of these categories. However, I am not sure where the mermaid on the Titanic fits in.

Four other observations with regards to Mrs. Waters' science classroom: (1) the lack of a sink with running water; (2) a lack of secure storage areas for lab equipment; (3) the location of the phone across the classroom from her desk; and, (3) the very small air conditioning vent high up on the west wall just above Doc Brown's DeLorean. This air vent is at the end of the line for the air conditioning ductwork, which transcends the length of the building's entire second floor. It passes through Rooms 208, 207, 206, decreasing in cross-sectional area until it finally reaches Mrs. Waters' Room 205. Even a mouse would find the size of the air vent claustrophobic! The force of the air at this endpoint is negligible, a mere whisper.

HELPING MRS. WATERS MAKE IT HER OWN

As mentioned in the preface, many of my decisions as to the extent and intensity of my roles as both participant and observer relied heavily on the definitions, recommendations, concerns, and cited works provided in Jackson's *Life in Classrooms*; Danny L. Jorgensen's *Participant Observation: A Methodology for Human Studies*, McCall and Simmons' *Issues in Participant Observation*; and Eisenhart and Borko's *Designing Classroom Research*.

During my 79 days of active participation and observation in Mrs. Waters' classroom, several opportunities arose for suggesting and implementing changes to the physical classroom environment. In the remainder of this chapter, I present a few of the events that were related to these changes, then discuss life in Mrs. Waters' village.

When Mike Skeldon suggested that Mrs. Waters make her class her own, this posed somewhat of a dilemma for me. If the opportunity presents itself, should I suggest some physical changes to the classroom? And, if so, of what sort? Will Mrs. Waters accept my suggestions, or is she is likely to not exactly agree and just be nice by compromising? One advantage of my participatory role on the "complete outsider"–"complete insider" continuum was that changes could be made in real-time. They did not have to wait to be put into place the following school year.

However, I believe that any suggested changes would have to be handled as discretely as possible (i.e., behind the scenes) so as not to influence student perception of me as a fellow student. They needed to accept me.

A CAREER TIP

One should always be congenial to everyone at work, but there are two people who deserve particular attention: your boss and the maintenance supervisor. Where the boss is concerned, this is common sense. However, most employees in businesses and professional organizations underestimate the importance of the maintenance department. In fact, these personnel generally have the keys to every part of the building; also, they are often sources of great information as they overhear conversations while performing their duties. Furthermore, they know where to get hold of things that are not in immediate view, but can be found with a little scrounging around. For a teacher, this can mean access to a better desk, a more comfortable chair, a filing cabinet, some paint, or even a badly needed bookcase. In both my careers as an engineer and teacher, I have found that making friends with the maintenance department has paid off.

Perhaps this was the problem in the Fogarty Building. In the two years I taught there, I never saw a maintenance person—I wonder if we even had one. Certainly this would explain the lack of toilet paper, paper towels, and hand-soap in the bathrooms.

So, it is always great when Brian, Beacon's maintenance supervisor (my new best friend!) informs me about items that have arrived in the building from donors or auctions and are as yet unclaimed. Incidentally, he doesn't like being referred to as the "janitor"—he tells me he prefers the name "maintenance god."

Wednesday, March 9, 2011 (Day 19). Brian informs me that he has just moved several choice items from the old Woonsocket Middle School's recently held auction into temporary storage on the fourth floor.

Timing is everything for a successful hoarder! I quickly make my way to the fourth-floor storage space behind the film classroom where I find a treasure trove of miscellaneous furniture. Of particular interest are four science lab carts in various condition, along with one small and three large, heavy duty, metal locking safety cabinets—the kind used for storing hazardous materials (e.g., corrosive, combustible, flammable). This offers some possibility for solving at least three of the physical environment issues I noted when I first entered Mrs. Waters' classroom. At least one of the lab carts has a sink configured to work with a pump and a tank under the sink. I see one hand-pump, but the tank is missing. The cabinets will certainly clean up all of the lab equipment currently scattered around Mrs. Waters' room. In particular, removing the equipment stored on the table along the south wall will allow the table to be gotten rid of altogether (see "S" in fig. 1.2) and so free up ample space for students and Mrs. Waters to walk behind the seats of those students that sit along the south wall. I have to act quickly to claim my prizes.

Thursday, March 10, 2011 (Day 20), 7:00 a.m. Back up on the fourth floor with my tape measure, I jot down the dimensions of the four cabinets. The three large ones are forty-three inches wide and the short one, thirty-five. This makes for a total required wall length of 164 inches if the four cabinets are placed adjacent to each other. I therefore proceed to Room 205. On the north wall, to the left of Mrs. Waters' desk, I measure the length of the wall 189 inches. Perfect! There is indeed sufficient space for them. Well, almost perfect. There's an electric outlet along the wall that will need to be accessed regularly. A gap will have to be left between two of the cabinets. However, the cabinets themselves will need some work before they can be brought down to the classroom. Each one is painted a different color. It would be nice if they were uniform. I'll have to get Mrs. Waters' opinion on this. Also, there are a few dents that will probably need to be hammered out. And, the adjustable metal shelves inside are missing those small support clips that pop into slots inside the cabinet and allow the shelves to be mounted at various heights.

8:00 a.m. I decided to visit Mrs. Waters. It's her free period now and I can have a one-on-one conversation with her since her students will not yet have arrived.

Today is lab day and she is setting up her stations. She has chosen Thursdays for this since first period was her free and so allows her time to rearrange the classroom tables to form the work stations. For this particular lab, Mrs. Waters has rearranged the tables into four stations, each consisting of two tables back-to-back with two chairs on both sides. They are labeled workstations one through four.

I spoke with her about the lab carts I had located. She said she only needed one and showed me a nice one that had a sink. It was located in the other science room (204) just next door and just happened to match her projector cart. Since Room 204 already had a working sink mounted against the wall, there was no need for a lab cart with a sink. Penny [pseudonym], the substitute science teacher, had placed a board over the sink on her lab cart and was using it as a projector cart.

I decided to swap out Penny's lab cart in Room 204 for one of the auction carts. I also recall noticing that one of the four auction lab carts was designed without a sink. That one would be perfect for Penny as no board would be required over the sink. When I explained my plans for the cart swap to her and noted that Mrs. Waters needs a sink, she thought it would be great. She liked the idea of having a cart with a solid top rather than one with a sink. Under the current set up, her projector could fall into the sink! I left Penny feeling really proud of my successful bartering.

Friday, March 11, 2011 (Day 21), 7:55 a.m. Arriving at class, I help move all of the tables back into the U-shaped configuration from the lab station positions of the previous day. What a lot of confusion there was as to the original configuration of the tables! Fortunately, I was able to pull out my sketch from my field notes from Day 1. Later, these served as the basis for figure 1.2 as presented earlier in this chapter.

Thursday, March 17, 2011 (Day 25), 8:25 a.m. Today is another lab day. This time Mrs. Waters has moved some classroom tables up against the wall along Church Street in order to put together the required lab stations. This particular lab, called the "Gold Coin Lab" requires the use of electric burners and she doesn't want any electric cords on the floor—someone could trip on them. Safety is always a prime consideration for her. I make a note in my journal that that moving of tables to form lab workstations and then the repositioning of them back to the regular "U" shape configuration is a nuisance.

Friday, March 18, 2011 (Day 26), 7:45 a.m. Mrs. Waters tells me that Ashley, who had been absent yesterday, wanted to make up the "Gold Coin Lab" we had just completed, but Mrs. Waters couldn't have her do it. The amount of time required to both set up for, and clean up after, this type of hazardous material lab is excessive when just one student is involved. She explains to me that this is a commonly occurring problem.

I suggest to Mrs. Waters that we mount a camcorder on the newly acquired lab cart to film, project, and record the students' experiments. This would give us the ability to archive the filmed experiments on her Digication[3] site. Students who had been absent the day of the lab would then have access to this information on film. She thought this was a good idea. I offered to do some research to see if this could be done, if not for this semester, perhaps for next year.

Wednesday, March 23, 2011 (Day 29), 11:30 a.m. In Chemistry Block 1, the phone rang in the classroom. It was Brian. He was going to Lowe's and wanted to know what I need to modify the lab cart and what color the paint should be for the four cabinets. I ask him to look for a pump and a container for under the sink. Mrs. Waters decides that she would like all four cabinets to be painted gloss black.

Friday, March 25, 2011 (Day 31), 7:55 a.m. I am usually one of the first students to arrive in chemistry class and in my seat before the others who sit next to me—normally Sima, Sarah, and Manar. When I see them enter the room, I get out of my seat so they can pass between the back of my seat and the table of lab equipment that abuts the wall behind our chairs. (see "S" in fig. 1.2).

Once again, I make a note to myself that I need to arrange Mrs. Waters' room to eliminate the table behind me—we're more than thirty days into this and I haven't solved this problem yet! I know the solution is to move those four cabinets down and stock them with all of the lab equipment that is now in various locations around the classroom, especially those on the table behind me. I need to have Brian get me the paint and supplies ASAP—I proceed to hunt him down. He tells me he will get them next week. I express my concern about the glossy finish that is presently on the cabinets and ask whether he thought I should sand them first. I would hate to put all of the effort into painting them only to find the new black gloss paint peeling off in sheets. He suggested that I just rough up the surfaces with an electric palm sander. He would loan me one. Then I could wipe the surfaces with some alcohol pads before putting on the paint.

Friday, April 8, 2011 (Day 41), 7:20 a.m. I sign in at the office and proceed down the hallway to the elevator. I am considering checking out the painting supplies that Brian had purchased for Mrs. Waters' cabinets. As I near the elevator, I hear a female voice screaming from behind the elevator doors and somewhere up the shaft. I believe it's Penny, our substitute science teacher. Immediately, I run up the back stairwell to the second floor, but the screaming voice is still further up. So, I go to the third floor. Penny is still somewhere further above me. However, her voice is louder and becoming ever more distinct. She is definitely in some kind of pain: "Help, somebody help, I've got to go to the bathroom! I can't get the door open."

By now, I have arrived at the fourth-floor elevator doors. In spite of the tremendous upper body strength I've been blessed with and all the might I am able to muster, I am unsuccessful at prying them open. Meanwhile, the noise has drawn the attention of humanities teacher Jen Wall. She comes running over and is now on her cell phone calling for help. I quickly back away from the elevator and proceed to my office as there is really nothing more I can do—I'm thinking Jen will handle it.

I have heard of cases where extremely heavy objects, like cars on jacks, have fallen on a worker beneath them. Someone nearby comes to the rescue and is somehow able to instantaneously muster superhuman strength and lift the car, thereby freeing the victim.

Penny must have had to urinate very badly because she herself was eventually able to force the doors open from the inside before help arrived. The elevator had stopped about a foot short of the fourth floor so it was just a short climb for her to get out of the elevator. I'm glad it all worked out. And admittedly, I did get a little needed exercise running up the stairs. I made a note that this was my first experience as a *first-on-the-scene responder*. Now, back in my office, I'm thinking that because of this distraction, checking out the painting supplies will have to wait until after school.

Recall that one of my goals was to be very discrete with regard to changing the classroom's physical environment. I wanted to maintain the students' perception of me as just one of them. Painting the cabinets would be a major undertaking, messy and smelly. It could not be done during the school day. Doing it before the start of morning classes was out of the question because of the mess it would likely make. And, there would be limited time for set up and cleanup. I also did not relish the thought of walking around the school during the day reeking of paint remover. After school work on this project would also be problematic because I had meetings or club activities just about every day. It would be very late in the day before I could begin painting. So, all of these considerations taken together, if I were to protect my role as participant observer, using vacation time to paint the cabinets was really my only option.

APRIL VACATION WEEK

Public schools typically shut down for one week in April—it's tradition. Beacon Charter High School is no exception. Given that nobody would be around all week to provide distractions, I expected to get a lot done.

Tuesday, April 19, 2011 (Vacation), 8:00 a.m.–4:30 p.m. There was nobody in school today; so, no distractions. I put on some old clothes and work on sanding down the cabinets on the fourth floor. There are large stickers permanently glued to the metal doors that can only be removed in small bits with a scraping tool. The work is tedious, messy, and boring—and this is just the beginning of a long process.

I worked on the outside of the first tall cabinet to the point where I had completed the application of the first coat of paint. It was clear that it would take at least two coats and a full day between them for drying. So, I went down to Mrs. Waters' classroom to start reorganizing things.

By the end of the day, I had put in eight and a half hours at school.

Wednesday, April 20, 2011 (Vacation), 7:00 a.m.–9:00 a.m. I finish painting the inside of the first cabinet and apply a second coat to the outside surfaces I had previously painted. Then I stop and clean up to prepare for a visitor:

9:00 a.m.–7:00 p.m. Carolyn Taylor, our Special Education Director, had called me last night to see if I was going to work at school today. I explained that I intended to work on the cabinets for Mrs. Waters' room. She asked me if I could spare a couple of hours in the morning to help her work on two deliverables for the school: An RTI (Response to Intervention)[4] Self Assessment" and the school's "Guiding Principles for RTI." Carolyn and I both started at Beacon in 2005 and have worked on many projects together. The problem is that many of Carolyn's projects, which are generally always very worthy endeavors, often entail hours and hours of uncompensated time. So, we agreed to meet today at school for a couple of hours from 9:00 a.m. to 11:00 a.m. Not unexpectedly, Carolyn's two-hour window ended up amounting to eight hours of work. She left around 5:00 p.m. and I dragged myself up to the fourth floor. Needless to say, I was too tired to do a lot of painting at that point. I worked for a couple of hours and then drove home. By the end of Vacation Day 3, I had put in another ten hours at school—only two of which were spent on the Room 205 project. Thanks Carolyn!

Thursday, April 21, 2011 (Vacation), 8:00 a.m.–7:00 p.m. The pressure is building! I am more than halfway through my vacation week and there is so much more to do before everyone comes back to school on Monday. I decide to concentrate on reorganizing Mrs. Waters' room. Painting the cabinets upstairs would have to wait. I needed to paint the north wall where the new cabinets would be positioned. I had previously gotten some white paint from Brian for that purpose. This meant painting over the two dinosaurs – alas, a second extinction!

Once the wall was painted and in the process of drying, I went looking for a hand truck to move the first tall, finished cabinet down into the room. There was one sitting on the back loading dock. There were also four brand-new small folding tables leaning up against the wall. These would be perfect for science lab station tables. Mrs. Waters wouldn't have to re-arrange her classroom tables on lab days. I did hear a rumor that Jason (Jay) LeClair, one of Beacon's art teachers, wanted to use them for his classes—this usually meant certain death for tables of any sort! I am going to do my part for society and rescue them, sort of like adopting a dog or cat from an animal shelter. Sorry, Jay!

With the hand truck, I moved the four folding tables to Room 205, then the tall shiny black, gloss cabinet, the first one to be painted. I positioned the cabinet along the freshly painted north wall just to the left of Mrs. Waters' desk—it looked great! However, having spent eleven hours on Vacation Day 4 in school—I was now tired!

Friday, April 22, 2011 (Vacation), 8:00 a.m.–7:00 p.m. I worked another eleven hours today, the last day of the vacation week. Time is definitely running out—I realized I would not be able to finish painting the next two cabinets and get them positioned for the return of students on Monday. So, again I decided to concentrate on getting all of the other things done that needed doing so that Mrs. Waters could be ready to make use of the classroom on Monday without any obstructions.

I rebuilt the large wooden bookshelf that had completely fallen apart because various heavy books completely had pushed out its backing and stripped its screws. I went to Lowe's and purchased some good nails and was able to completely secure the backing and all the shelves—it won't fall apart in the next decade. I restocked the shelves in a very organized manner with the books I had removed prior to making the repairs. I also repaired a smaller wooden bookcase that was falling apart.

There was a big oval table positioned in the front of the room near the whiteboard. It was constantly obstructing the path of everyone who walked by (see "M" in fig. 1.2). Mrs. Waters used it mostly for handouts and board markers. I moved it to the north side of the column, to the left of the classroom tables. Best of all, I placed a nice 28-slot mailbox unit on it against the backside of the column (perfect for organizing Mrs. Waters handouts). Positioning the oval table and mailbox required that I remove and reposition two plastic magazine holders that were screwed to the column. The mailbox unit had not been claimed by anyone because it had been stored on its side, and the partitions (i.e., shelves) were not in place, so its functionality was masked. Everyone thought it was a broken bookcase that was missing shelves. In fact, I overheard Principal Pilkington a few days ago saying that he was surprised nobody had taken it. Within minutes of his disappearing around the corner, the mailbox was in my possession and on its seventy-five foot journey to its new home in Mrs. Waters' classroom. What a great heist—and a cause of jealousy on the part of some other faculty members!

I cleaned up the black metal bookcase that was placed up on top of a small, wooden rectangular table tucked in the southwest corner of the room. The latter was used to store chemistry and physics textbooks. I removed the table and placed the bookcase on the floor up against the south wall. This allowed me to reposition the small table by the whiteboard where the oval table had been originally placed. The black cabinet was missing four shelf-mounting brackets—one shelf was unusable altogether. I searched all the other classrooms to see if there were any similar bookcases that had extra brackets. No luck. I went to Lowe's in search of them. Those that I found there didn't work out. I would have to make something special—yet another item for the "to-do" list.

I removed the lab equipment on the large, awkward double-decker computer table behind my seat and stocked the items in the tall metal cabinet I had recently painted. The odd-shaped computer table was originally Skeldon's, then I had claimed it, now it was Mrs. Waters.' It had blocked everyone from passing behind my chair and forced them to walk around the other side of the classroom. For a few of them, this translated into walking close to an extra mile over the course of a semester—not a bad strategy if some were over-weight! But moving this table also freed up a lot of extra space in the classroom.

I moved the actual computer table, which contained a computer and printer, to the left side as you enter the classroom. There is a small indented area where the north wall juts in, just sufficient in length to accommodate the 72-inch long table.

On the thirty-minute drive home, I reflected on the 40-½ hours I had put in over the vacation week. I felt I had accomplished a lot and learned a number of lessons in the process. But, I had also ruined a good pair of sneakers on the first day of painting, and four pair of socks during the week—my best set of old clothes. I spent a few more hours removing the numerous small specs of gloss-black paint that had adhered to the lenses of my glasses as a result of the spraying. The safety goggles that I put on later probably saved a good bit of aggravation. I decide that the many hours of work in school during vacation week deserved a trip to Taco Bell and, also, that it gave me license to take both Saturday and Sunday off.

I was worried about Mrs. Waters' reaction. Maybe I was intruding on Skeldon's recommendation to her, i.e., that she "make the room her own"? She already had a very nice classroom—it just needed some tweaking. And, from my observations thus far, her arrangement of the classroom tables in the U-shaped pattern was ideal for teaching and learning. Her desk was always neat and clean and in the perfect position in the classroom. However, the lab equipment was scattered around the room due to insufficient storage space; the placement of the oval table by the whiteboard and the odd-shaped table behind my seat both obstructed the flow of traffic through the room; the wooden bookcases were in need of repair; etc., and so represented further opportunities for improving the physical environment of the classroom. I began to make a mental list of what I perceived to be the improvements I had made for her:

- She definitely had more space.
- She now had four nice mini-tables to use as lab stations, thus eliminating, or at least minimizing, the need to reposition classroom tables for lab day.

- She has a nice mailbox organizer (the envy of many of her peers).
- She would end up with four heavy-duty locking metal storage cabinets for organizing her lab equipment and supplies.
- The wall behind the cabinets had gotten a badly needed fresh coat of paint.
- Two bookcases were rebuilt.

And, my thought process continued:

If I hadn't hoarded the four metal storage cabinets, the mailbox unit, and the four brand new, small folding tables, they might not have ended up in Mrs. Waters' room. Who would have painted the metal cabinets if I hadn't? Who would have repaired the bookcases?

Just before I pull into the parking space at my apartment complex, I question whether I have broken any of the rules of participant observation research as they were laid out. I'll need to revisit Jorgensen's book.

Monday, April 25, 2011 (Day 47—First day back from Vacation) 7:55 a.m. I proceed to chemistry class. Mrs. Waters is pleased with all the work that has been done. During class I write in my field notes: "Today was the first time in 47 classes that Mrs. Waters walked along the wall behind my seat—I had finally gotten rid of the table from hell."

Prior to this, I also noted that no student in any of Mrs. Waters' classes in Room 205 had complained about, or made suggestions on clearing the obstructed pathway. In my experience, students rarely advocate for the betterment of their own learning experience where the physical classroom environment is concerned. Instead, they often become complacent and just quietly adjust to less than desirable conditions—lighting, temperature, noise, working space, furniture, etc. However, all of these things can affect their engagement with the subject matter and, ultimately, their achievement. Many professional conferences often ask session attendees to complete surveys. In addition to questions regarding the effectiveness of the speaker and the quality of the content of presentations, most of these surveys include at least one question on the facilities (i.e., the physical environment). It should be no different for our students—we should welcome student feedback on this and other issues as a means of improving teaching and learning. In later chapters, I provide examples of how Mrs. Waters and I used student surveys to inform the process of instruction.

Tuesday, May 3, 2011 (Day 53), 7:55 a.m. I get a call from Mrs. Waters thanking me for the cabinets. Finally, all four were done! All with a shiny black gloss finish against the clean, freshly painted white wall. It looked great! She said that she would like to move the oval table from its current position (adjacent to the column) to the area now occupied by the fish tanks. I told her that this seemed to me a good idea. She also told me that she was able to get rid of her old, grey metal cabinet that she had been using to store glassware. Also, she said she wished to get rid of two short bookcases. I was feeling really good about our collaboration toward these physical changes and she was indeed making the classroom her own—with a little help from a participant observer.

Monday, May 9, 2011 (Day 57), 9:40 a.m. The Chemistry Block 1 class is ending. Sarah, two seats to my right, cuts herself on the sheet metal baseboard unit behind her after getting up from her seat. The metal housing had been bent out of shape since the beginning of the year, and probably long before that. It had gone unnoticed because the odd-shaped computer table had blocked the view of it. But this was now gone. Fortunately, the injury was very minor—it could have been much worse. Sheet metal can be very sharp. I'll have to get Brian to fix this right away.

Wednesday, June 8, 2011 (Day 78), 11:15 a.m. Today our chemistry class was moved to Room 303 (Peter Boland's English classroom) because interactive whiteboards[5] are being installed in the math and science rooms. The change of venue is good because it's over 90 degrees in Mrs. Waters' room on this June day. She has that very small air conditioning vent, with the

Figure 1.3. Mrs. Waters' new interactive whiteboard.

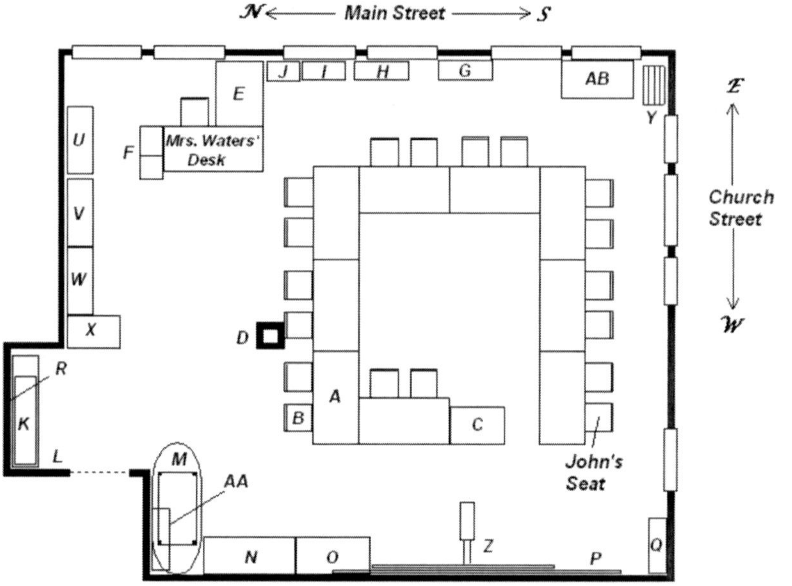

A. Classroom table: 60" x 30"	***N***. Computer table w/two computers and one printer
B. Chair	***O***. Table: 48" x 24"
C. Old projector table	***P***. Standard whiteboard
D. Column	***Q***. Black metal bookcase for textbooks
E. Mrs. Waters' desk	***R***. Bulletin board
F. 2-Draw filing cabinets	***U***. New tall metal storage cabinet #1
G. Tall wooden bookcase	***V***. New tall metal storage cabinet #2
H. Metal bookcase	***W***. New tall metal storage cabinet #3
I. Short wooden bookcase	***X***. New short metal storage cabinet #4
J. Chrome metal shelf unit	***Y***. Stacked folding tables: 48" x 24"
K. Fish tanks and table	***Z***. Interactive whiteboard w/ceiling mounted projector
L. Phone jack	***AA***. 28-Slot mailbox unit
M. Large oval shaped table	***AB***. Mobile lab cart w/sink

Figure 1.4. Mrs. Waters' classroom (Day 79).

faint whisper of assumed cold air, located up above the whiteboard (by Doc Brown's DeLorean). The classroom also has large windows on both the east and south walls. They are constantly admitting the sun's rays during most of the school year. All of these factors can make Room 205 very uncomfortable in June.

Mrs. Waters' newly installed interactive whiteboard and ceiling mounted projector (along with a glimpse of the DeLorean and mini air conditioning vent) are shown in figure 1.3.

Thursday, June 9, 2011 (Day 79—the last day of chemistry class). The physical changes made to Room 205 over the course of the semester are reflected in figure 1.4.

Unlike eight years ago (almost exactly to the day), there is no need for caution tape to block students or faculty from entering Room 205. The natives do not have to be moved to the Stadium Theatre across Main Street because of unsafe or unsanitary conditions like those experienced in the closing weeks of the 2002–2003 school year in the Fogarty building.

The effect of the physical classroom environment on teaching and learning is a research area in the field of education that has not been given sufficient attention. Based on my observations in Mrs. Waters' classroom, physical changes that improve efficiency can add valuable instructional time, decrease distraction, cause students to develop more consistent routines, and provide a model of structure (i.e., organization) for them; such changes can also help them to respect their learning environment.

CLOSING REMARKS FOR CHAPTER 1

To the best of my knowledge, none of my fellow students in Chemistry Block 1 were aware of the work I did behind the scenes to effect physical changes in their environment. As far as they knew, it was Mrs. Waters and Brian who were doing this. Certainly, none of them ever questioned me or made comment on it. All of my work had been done in the morning, after school, on weekends, and during my vacation time when they were not around. This was a good thing because I did not want them to see me as anything other than a fellow student. However, I will have to survey them after all of my research is complete (i.e., after the semester ends) to see how they perceived my presence in the classroom (see ch. 10 "Reflections on the Research").

NOTES

1. IEP (Individualized Educational Plan). Mandated by the Individuals with Disabilities Education Act (IDEA). Through an evaluation process, an IEP must be tailored to meet the special educational needs of a specific child, who may have a disability, as this is defined by federal regulations. The IEP is designed to help teachers and related service providers, e.g., paraprofessional educators, understand the student's disability and know how it affects the learning process.

2. Third daughter Seraphina was born November 18, 2011. This was after my 79 days spent in Mrs. Waters' Chemistry Block 1 class, but prior to my interview with husband Chris.

3. *Digication* is web-based software that provides a means of creating an online community that provides efficient communication between members. All of Beacon's students and teachers have their own personal Digication site. Beacon teachers post their syllabi, assignments, and resources. Students use it for archiving their work for their graduation digital portfolios (i.e., e-portfolio).

4. RTI (Response to Intervention) is a systematic process in education designed to detect and assist children who are having difficulty learning. It involves early intervention, frequent progress measurement, and increasingly intensive research-based instructional interventions for children who continue to have difficulty. Children who do not respond to effective RTI interventions may have diagnosable learning disabilities and warrant special education services.

5. Interactive whiteboards are used in conjunction with a computer and a projector. The projector, often ceiling mounted, displays the screen from the computer's desktop onto the board. Electronic sensors built into the surface of the board are designed to sense the actions of a finger, pen, or stylus and so transmit data to the computer. Unlike the traditional blackboard and chalk, the interactive whiteboard/computer can capture, save, and print any images that are created.

Chapter Two

Going Native

My First Ten Days in Chemistry

Mike Skeldon, Academic Dean, says I won't last ten days. Dr. Pilkington, Principal, has given me thirty. With support like this from Beacon's administration on my planned research endeavor, admittedly, I'm feeling a little anxious about my upcoming first day in chemistry class. After all, there is a precedent for this.

A FIRST-DAY REFLECTION

October 6, 1977. This was my first day in ME 313 (Mechanical Engineering Lab I) at Western New England College (WNEC) located in Springfield, Massachusetts—since renamed Western New England University. It was the beginning of my junior year, but more notably, the beginning of my experience at the college. I had just transferred to WNEC after earning an Associate of Science in Engineering degree from Rhode Island Junior College (RIJC) located in Warwick, Rhode Island—since renamed the Community College of Rhode Island. The name of the junior college was changed in 1980 because RIJC students were increasingly being referred to as *Rejects* because they were perceived as not being able to get into a four-year college or university immediately out of high school.

ME 313 was the first in a four-semester sequence of lab courses that were required for all mechanical engineering students. This course involved a different lab experiment each week; one week following the lab we were to submit a formal report on our research. The lab consisted of small groups of three or four students. These were assigned to one of the four mechanical engineering professors. Each professor was responsible for one of the different lab stations which were distributed across the fairly large mechanical engineering laboratory in Sleith Hall.

Today, I was assigned to Group A with lab partners Andrew J. Cobb, Alan C. Dean, and Richard B. Hossack (Who remembers their lab partners 34 years later?). This was the first time I had met any of them, as I was a new transfer student. All three of them had entered WNEC as freshmen. Dr. Wellen G. Davison (acting Dean of the School of Engineering) would be instructing us on the Electric Motor lab experiment. My first impression of Dr. Davison was that he was going to be tough. He had a very stern expression and a *strictly business* manner.

The electric motor lab station was set up on the far side of the mechanical engineering laboratory. The experiment involved investigation of an AC (alternating-current), one-quarter horsepower, three-phase motor. Standing by the motor, Dr. Davison began his lecture. The four of us were standing around him—more exactly, I was unconsciously leaning up against a table behind us.

All of a sudden, Dr. Davison pointed his finger at me and yelled loud enough to get the attention of the three other students in my group, the twelve other students in our class, the three other professors, the mechanical lab supervisor, and most likely the secretary in the engineering department office down the hall:

"That's just what I don't want you to do! Don't lean on the equipment. If you don't know what to do with your hands, put them in your pocket!"

Totally embarrassed and shocked by his reaction, I shoved my hands into my pants pockets so fast and so deep that I felt and heard the fabric tear. Dr. Davison continued with the experiment.

About, twenty minutes later, suddenly, Dr. Davison jumped a foot into the air, violently waving a finger a few times and then wetting it in his mouth. Apparently, he had just severely burned it on the very hot metal housing on the motor. Without any pre-meditation, I immediately and uncontrollably blurted out:

"See if you had your hands in your pocket, that never would have happened!"

It was too late; I couldn't take the words back. What the heck did I just say? Dr. Davison looked at me, broke his stern look, cracked a little smile and simply said, "You're right."

He could have immediately thrown me out of class or even out of the entire engineering program. But he didn't. He gave me a chance. However, a week later, I did receive a "50" for a grade on my lab report—not as retribution for my remark but because I made several mistakes in the write-up. Over the following two years, I developed a great respect for Dr. Davison. I eventually earned some regional and national recognition for the college after having successfully competed in several engineering competitions. And, Dr. Davison contributed significantly to my success. As concerns his grade of "50," this would be the first and last time I ever received a failing grade on any report, or writing assignment, or in any content area, for that matter—ever! Lastly, as a teacher, I have had several incidents over the years in which students have blurted out something or behaved in a less than desirable fashion in a moment of frustration. So, I always reflect back on Dr. Davison's controlled response and on the fact that he gave me a chance to succeed.

However, this time around as a student, I have made the firm decision not to take any chances and to restrain myself from blurting out any remarks should Mrs. Waters burn her finger in chemistry!

THE NEW KID IN CLASS

Thursday, February 3, 2011 (The day before my first day). Everything seems set to go. Mike Skeldon (Academic Dean and Principal in Residence) and Dr. Pilkington (Principal) have placed their bets. Robin Murphy (Dean of Students and Guidance) has officially scheduled me into PowerSchool, a widely used web-based student information system that Beacon uses. As shown in figure 2.1, I have received a letter with instructions on how to access my grades on Beacon's PowerSchool. All of Beacon's students have received a similar letter with their own personal username and password to access their grades.

MY SCHEDULE

I have created a table for my Chemistry Block 1 class schedule (see table 2.1). Beacon uses a schedule with four blocks per day. These rotate through the week. For Monday: B1, B2, B3, then B4; Tuesday, B2, B3, B4, then B1; Wednesday, B3, B4, B1, then B2; Thursday, B4, B1, B2, then B3; and, Friday, the schedule is the same as Monday.

Mondays and Fridays begin with a half hour of "Drop Everything and Read," referred to as DEAR (to be discussed later). I plan on attending DEAR in Mrs. Waters' Room 205 as often as possible since this is the same group of students that will be in my Chemistry Block 1 class and there is no change of classrooms between DEAR and the start of chemistry class. Also of note is that Wednesday's class occurs during the lunch block (11:15 a.m.–1:05 p.m.) Unfortunately, we will be going to the last lunch. It starts at 12:45. Waiting that long to eat may be my biggest challenge!

Beacon Charter High School for the Arts
320 Main Street
Woonsocket, Rhode Island 02895

Dear John Butler:

Timely access to your grades is important to your success in school. To help facilitate this access, we are providing you with a student login and password for your online grade book, PowerSchool.

As of right now, PowerSchool includes information for the current school year. We are in the process of importing information from prior years. Please have patience while this process continues.

This information should help you stay on top of your progress.

If you have questions for your teachers, you should only bring them to your teachers' attention at the appropriate time. This usually means after school or before school and NOT during class. Please be respectful of your teachers as well as your fellow students by speaking to your teachers outside of scheduled class time.

In order to access your grades, please visit:

Beaconart.powerschool.com/public

And enter:
USER NAME: ****
PASSWORD: ***

If you have any questions or concerns regarding access, please let me know.

Regards,

Michael Skeldon
Academic Dean

Telephone 401.671.6261 • Fax 401.671.6264 • www.beaconart.org

The Beacon Charter School does not discriminate on the basis of age, color, sex, sexual orientation, race, religion, national origin, or disability.

Figure 2.1. Instructions for student's online access to grades.

Table 2.1. My Chemistry Class Schedule

	Day of the Week				
Time	*Monday*	*Tuesday*	*Wednesday*	*Thursday*	*Friday*
8:00–8:30	DEAR				DEAR
8:30–9:45	Chemistry				Chemistry
9:50–11:10				Chemistry	
11:15–1:05			Chemistry		
1:10–2:30		Chemistry			

THE SEMESTER BEGINS

Friday, February 6, 2011 (Day 1). Today should have been my first day in chemistry class with students. However, since Mrs. Waters will be incorporating a state-standards-web-based science instructional tool, Study Island, into her curriculum, she will be participating in a webinar today for professional development (as noted in the preface). Joining her will be Carolyn Taylor (Director of Special Education); Larissa Perkins (math teacher), who will be using the math component of the software in her classes; and myself. Bill Connell, a special education teacher, will cover Mrs. Waters' chemistry class.

We have been using Study Island for a few years now in our math classes so students are already familiar with the program. This year, one difference is that the company, Archipelago Learning, has changed the name of its Study Island product to "The Island" to differentiate the high school level version from others. However, since our students have become familiar with it by its former name, I will continue to refer to it as Study Island.

The delay in starting chemistry today is not a problem. In fact, there's something awkward about starting a course on a Friday.

Monday, February 7, 2011 (Day 2), 7:00 a.m. I arrive at school, sign in at the main office, and proceed to my office on the fourth floor. Most of our classes are semesterized—they meet every day for one-half of the school year. This was the day I would officially begin chemistry class as a student. I had chosen Mrs. Waters' eleventh-grade Chemistry Block 1 class for my *participant-observation* study. It began on Mondays at 8:00 a.m. I dug out a backpack from behind several boxes that were stored in my office. I figure the backpack would help the students in chemistry think of me as a fellow student. The black backpack had a large circular New England Laborers' decal on it. It was a souvenir I had acquired two years previously while chaperoning for the annual "Construction Career Days" field trip. People at the school tend to regard me as a sort of a hoarder—perhaps the only explanation for why, two years later, I still had the backpack. I packed it with a wide-ruled

70-sheet spiral notebook, a scientific calculator, a 12-inch clear plastic ruler, a pencil box filled with several sharpened pencils, miscellaneous colored pens, a yellow highlighter and a white-out pen, a pack of sticky-notes, a brand new 200-page journal that I would use for my field notes; and, my personal *Student's Agenda.* All students are given a personal planner at the beginning of the school year. Referred to as an *Agenda,* it contains calendars and academic resources. The agenda also serves as a place for recording homework assignments and a hallway pass—to be signed by the teacher if the student needs to leave the classroom (e.g., bathroom, main office, nurse).

An *out-of-body* experience starts to happen for me! I am slipping away from being a teacher, and morphing into my role as a *student.* I decide to try on my new backpack. Did you ever fly on an airplane and find that the seatbelt was adjusted very short because a three- or four-year old had been strapped in your seat on the previous flight? Well, I needed to adjust my backpack straps to gain a little more length to tolerate my upper body frame—a product of several years of bodybuilding. Even with the straps fully extended, I felt like I was having a pair of tourniquets applied to my shoulders.

7:50 a.m. With backpack on back, I enter Mrs. Waters' classroom, Room 205. Since my eyesight is diminishing now at around age fifty (I wear eyeglasses), I wanted to arrive early so as to get a seat close to the whiteboard. Over the many years spent in classrooms with unassigned seating, both as a student and teacher, I have observed that the first few days are the most critical. Humans are territorial in nature. It seems like, after just a few days of class, we come to own the seat we initially chose–and for the rest of the semester! Similar self-proclaimed entitlements occur in the teachers' parking lot where unassigned spaces are concerned. Thus, it was imperative that I get to class early before any student could claim prime real estate.

I'm the first student to arrive in Room 205. The classroom tables are arranged in a "U" configuration (ch. 1, fig 1.2). I head for the end-seat (refer to "B" in ch. 1, fig 1.2) on the "U" closest to the door and the board. I quickly position my backpack on the table and take my seat. Other students are now entering and marking their territory.

Nobody has been told about my research or plans or my going *native.* So, I'm not sure what the students are thinking as they see me sitting in their class. However, they all know me since I was their 9th-grade Algebra 1 teacher. One of them, Jessica P., sits down next to me and says "Good morning Dr. Butler."

I start to realize that I made a poor choice where seating was concerned (nothing to do with Jessica P.). It is close enough to see the writing on the board, but there is an irritating beam of light entering the window on the south wall directly across from my seat. I should have grabbed the chair directly opposite me. But it's too late, Sima has already claimed it.

8:00 a.m. Mrs. Waters reminds everyone that today is "Drop Everything and Read (DEAR)" day. At Beacon, the time between 8:00 and 8:30 on Mondays and Fridays is reserved for silent reading school-wide. The actual chemistry class will begin at 8:30 and end at 9:45. No students change classrooms between DEAR and the start of the content class. In theory, there is no loss of time due to the transition from DEAR to chemistry.

DEAR is a school-wide silent reading activity that was implemented a few years ago to help boost our state reading scores. It is now part of the Beacon culture. Students are instructed to bring a book or other reading material of their choosing to DEAR. Most teachers provide reading materials in their classrooms for those students who forget to bring them. Unfortunately, in my haste to pack my backpack, I neglected to include a book to read.

However, Mrs. Waters maintains a collection of science related books and magazines so I have chosen a *Scholastic Science World* magazine. The 14-issue magazine is published bi-weekly and features current news snippets, articles and hands-on activities that cover every area of the science curriculum including life, earth, physical sciences, environment, space, health, and technology.

I grab the February 14, 2011 issue (Volume 67, No. 9) and start thumbing through the pages. The format is attractive because there are a lot of very short articles. The first thing that catches my eye is a picture of a pair of lungs with an adjacent title in large type, "Lab-Grown Lungs." Apparently, a team of Yale scientists has been able to grow a fully functional set of rat lungs in the laboratory—an approach that someday could be used to replace human lungs that are failing.

Not being much of a cat lover, I pass over the next snippet entitled the "Purr-Fect Lick," although my eyes did catch the leading line "Why are cats daintier drinkers than dogs?"

Quickly, my eyes gravitate to a full-page photograph of a man strapped to a jetpack with the title "Blastoff!" It was in 1965 in a James Bond movie entitled *Thunderball* that I first saw a jetpack. In the movie, as I recall, James Bond straps on a jetpack, and fires the two jets which are aimed downward. Agent 007 is thereby lifted upward and over a building for a very short flight and so able to escape his pursuers. The one described in this article claims its two high-powered fans can lift a 240-pound man up to 8000 feet; apparently it flies at 60 miles per hour for thirty minutes on a five-gallon tank of regular gas. The 250 pound jetpack is commercially available at a price of $100,000; it is manufactured by the New Zealand based Martin Aircraft Company.

Since it is classified as an *ultra-light aircraft*, a person doesn't need a license to fly it—all that is necessary is that one be eighteen years old.

I really like this *ScienceWorld* magazine. It is a bit more engaging than the *Scholastic Math* magazine I have used in the past in some of my lower-level math classes. I did notice that my fellow classmates appeared involved in some sort of reading and the room was quiet enough for me to concentrate. I glanced at the clock and noted that DEAR was nearly over. In fact, Mrs. Waters was approaching the front of the class.

8:30 a.m. "Okay, DEAR is over, did anyone read anything interesting?" she asked. A few students raise their hands and briefly describe what they had read. I should be paying more attention but I am busy writing.

Before Mrs. Waters officially begins chemistry, she gives me a few minutes to explain to the class why I am sitting in class with a backpack. I had already spent a lot of time thinking about what I wanted to say:

> I'm going to be attending every chemistry class with you this semester. I am conducting some research for the school that requires me to participate in and observe Mrs. Waters' class. The research involves looking at teaching and learning from the students' perspective. So, I had Mrs. Murphy in Guidance sign me into this class as a student. I will be doing every assignment that Mrs. Waters gives us such as class work, quizzes, projects, tests, labs, and homework. I am asking you to call me "John" in this class and "Dr. Butler" outside of class.

Interesting enough, there were no discernible comments or reactions of any significance from the students. I suspect that because most of them had me as a math teacher in 9th grade, I was neither a stranger nor a threat to them. Or else, they thought I was some sort of nut or weird character because they had me as their math teacher and this news came as no surprise.

Previous to this first class, I had established four rules. These had not been shared with Mrs. Waters:

Rule no.1: When Mrs. Waters asks a question, I'll let everyone else in the class answer first (i.e., I'll never volunteer to answer first unless no one else steps forward).

Rule no. 2: If I see somebody breaking the rules in class, such as using a cell phone or cheating, I won't say anything—it's important that I be trusted and accepted by the students. I would, however, note these observations for future discussion. Exceptions to this would be those events that involve student safety or potential damage to school property.

Rule no. 3: I will refrain from cracking jokes, making comments without raising my hand and being called upon, or causing other distractions.

Rule no. 4: Based on my observations from both a student's and teacher's perspective, if I think of something that might help Mrs. Waters' instruction or classroom management, I'll discuss it with her after class. At no time will I correct, contradict, or interfere with her instruction or classroom management.

8:40 a.m. Mrs. Waters explains that she will have the daily activities and objectives written on the right side of the whiteboard for the start of each class and that we should check the board as soon as we enter the classroom. Today, she has written:

What will we be doing? 2/7/11
1) DEAR (Participation grade)
2) Go over course syllabus and class setup
3) Inquiry activity
4) Scientific misconceptions activity
5) Assign homework: Unit 1 slides #1–#16 (skip HW slide)
Objectives: Students will be able to
– understand what is expected of them
– investigate common misconceptions in science
– solve a puzzle that represents the process of scientific inquiry

Mrs. Waters passes out her syllabus. It is one page, copied front and back with a title:

Chemistry: 11th Grade 2011 Room 205.

The subheadings are: Instructor; Teacher Certification; Tutoring Hours; Course Description; Goals/objectives; Standards; Grading Policy; Late Policy; and Misconduct Consequences.

At the bottom of the back page, Mrs. Waters' has provided lines for both a parent/guardian and student signature. They are required to acknowledge that they have read the syllabus.

She reviews the entire syllabus with us, covering every item on front and back. She explains that we will be doing a "Chemical of the Day" and/or a "Problem of the Day" activity at the beginning of most classes. They will be written or projected on the board and we should start working on them as soon as we get to class.

In the syllabus, Mrs. Waters states that we will need three notebooks: one to be used for a lab journal, one for class notes, and another for warm-up problems (e.g., Chemical of the Day, Problem of the Day). We will not be using a textbook. She will supply us with all of the necessary resource materials.

Mrs. Waters points to her desk and tells us she has a folder marked "Block 1" for our class assignments when they have been completed—we are responsible for getting them into the folder. For homework tonight, we are instructed to have our parents review the syllabus and sign it. We are then to return it next class—a slight problem for me given that my parents are retired and living in Florida! I guess I'll have to fire up the fax machine tonight.

Next, we are handed a *Student Questionnaire* to fill out. Questions include requests for our email address, and information about our hobbies, and interests. Checking her agenda on the board, I see that the Inquiry Activity is next.

INQUIRY ACTIVITY

Mrs. Waters hands everyone a zip-lock bag with four polygonal-shaped paper puzzle pieces. "Take the pieces out of the bag. You have three minutes to solve it by forming a square," she instructs us.

Recall that Jessica P. is sitting next to me. As I quickly fumble with my four puzzle pieces—orienting them in every conceivable position (except the correct one), I am feeling rather uncomfortable. Without moving my head, and for a very brief moment, I notice out of the corner of my left eye that Jessica P. is done; and, she appears to have been given puzzle pieces of the same shape. That quick snapshot is enough for me to capture the image of the solution—a perfect square shape. I quickly maneuver my pieces into their correct position. Unfortunately, Jessica P. somehow catches my left eye and states rather loudly "John's cheating!" Okay, maybe I was. However, I would argue that the line between cheating and educational research is not always so clear-cut. After all, I'm supposed to be observing. Needless to say, I am embarrassed. What's worse is the fact that I've known Jessica P. for a few years now, having had her as a math student. And, I do love her dearly, as I do the other students in the class. But, she would be the first to admit that her problem-solving skills in my math class were less than exemplary. So, here I am a PhD copying from a high school student—not a good first day so far!

Next, we're given a fifth puzzle piece that is square in shape and has an "X" imprinted on it. We are instructed to reconfigure the five pieces we now have into a new square shape. Trust me, I had blinders on this time. After a few minutes, Mrs. Waters asked those who had solved the problem to help those who were struggling. Solutions to both puzzles are given in figure 2.2.

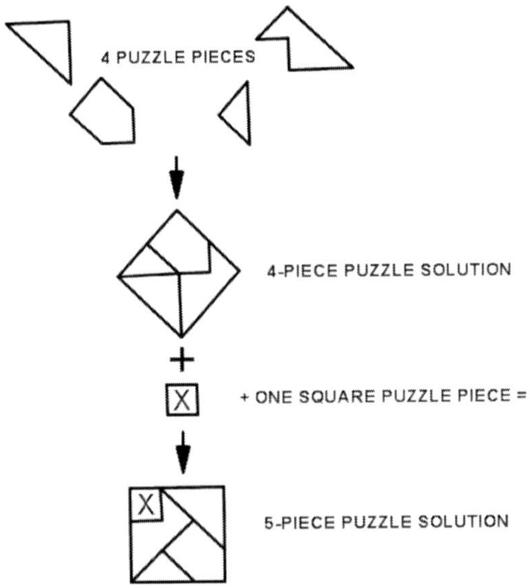

Figure 2.2. The puzzle activity.

Mrs. Waters states "We are surrounded by puzzles in nature. The puzzle-solving activity we had just completed required steps that define the process of *scientific inquiry*." She writes on the board:

Scientific Method and Inquiry:
(1) Observations of Nature generate questions = puzzle pieces
(2) Formulation of a hypothesis = the assembled pieces will form a square
(3) Test by experiment = using trial and error to rearrange the puzzle pieces
(4) Prediction = adding a fifth piece will also form a larger square
(5) Publishing results = so others can replicate the experiment and results

She then continues:

Nature, itself, is a giant puzzle, one that man has tried to understand and explain for thousands of years. Scientific method is the inquiry process that scientists use to help explain natural phenomena. More often than not, solutions are arrived at through the method of trial-and-error. In the case of Part 1 of this puzzle activity, there was no mathematical equation readily available for determining the position in which the four differently shaped puzzle pieces would form a square. The method of solving it involved first trying to match two of the pieces together, then adding a third, and then a fourth. Most of us had to make several attempts, mixing and matching them, before arriving at the solution. In Part 2, new information was introduced in the form of the additional puzzle piece. Hence, our solution to Part 1 was no longer viable for Part 2. We had to employ a trial-and-error method once again to arrive at a new solution, one that satisfied the condition of forming a larger square with exactly five pieces.

Mrs. Waters further explains that trial-and-error is generally a matter of luck but that it can be more systematic where there is collaboration.

Oh! What's going on here? I notice that Jessica P. appears to be texting on a phone that she is holding under the table in front of her. I ask her what she's doing and she shows me some game device with which she is playing. Unfortunately, my own Rule no. 2 (if I see somebody do something wrong, I won't say anything) precludes me from taking retribution for her "John's cheating" comment. However, I am definitely making a note in my journal.

THE SCIENTIFIC MISCONCEPTION ACTIVITIES

We were now handed a worksheet that had two different scientific misconception-related activities—one on the front side of the page and the other on the back. The first is entitled "Dihydrogen Monoxide"[1] whereas the second contains *true or false* questions entitled "Ten Scientific Misconceptions." Mrs. Waters asks us to read about dihydrogen monoxide and answer the question at the bottom of the page then turn the worksheet over and answer the ten misconception questions. The text on the front reads as follows:

> The chemical compound dihydrogen monoxide (or DHMO) has been implicated in the deaths of thousands of Americans every year, mainly through accidental ingestion. In gaseous form, it can cause severe burns. And, according to a new report, the dangers of this chemical do not end there. The chemical is so caustic that it "accelerates the corrosion and rusting of many metals, . . . is a major component of acid rain, [and] . . . has been found in excised tumors of terminal cancer patients." Symptoms of ingestion include excessive sweating and urination, and for those who have developed a dependency on DHMO, complete withdrawal means certain death. Yet the presence of the chemical has been confirmed in every river, stream, lake, and reservoir in America.

Judging from these facts, do you think dihydrogen monoxide should be banned?

As I am reading this, in my mind, I am picturing an open field cluttered with dented, lidless drums with skull and crossbones painted on the sides of them indicating that they contain toxic waste. So, I vote for a ban of dihydrogen monoxide.

After we have had a chance to read the front of the worksheet and answer the question, Mrs. Waters explains that the complicated appearing chemical name *dihydrogen monoxide* can be broken apart as follows: *di* means two and *mono* means one. Therefore, we have two hydrogen atoms and one oxygen atom:

$$2\ H + 1\ 0 = 2H10 = H_2O = \text{water}$$

So, this is just water! She tells us that although the name dihydrogen monoxide follows the accepted rules of chemical nomenclature, it is hardly ever used. She explains that this exercise is based on a hoax. It is commonly used in science education to emphasize the importance of developing sound scientific literacy.

"Mrs. Waters, can I go get a drink of dihydrogen monoxide?"

We now turn over the worksheet.

Here are the ten true or false misconception questions on the back of the worksheet (Mrs. Waters and I invite the reader to try to answer them). Answers are provided in the Note no. 2 at the end of this chapter:

1. We experience winter when we are furthest from the sun (T or F)?
2. Blood is blue when it is in our body, and it changes to red when it comes in contact with the air (T or F)?
3. There are nine planets in our solar system (T or F)?
4. Objects float in water if they are lighter (T or F)?
5. The yolk of the egg is where the embryo grows (T or F)?
6. Hair and fingernails grow even after a person dies (T or F)?
7. Daddy long legs is the most venomous spider in the world (T or F)?
8. Shaving causes hair to grow back thicker (T or F)?
9. Humans only use 10% of their brains (T or F)?
10. All cells come from existing cells (T or F)?

After about five minutes, Mrs. Waters stops us and we discuss the answers.[2]

9:42 a.m. Class is just about over, Mrs. Waters assigns the homework for tomorrow: In addition to getting three notebooks, we need to get the course syllabus signed by a parent or guardian. They need to acknowledge that they have reviewed Mrs. Waters' expectations and policies. Also, we need to go to Mrs. Waters' Digication page, select "11th Grade Science," click on "PowerPoint Notes," and then "Chem. Unit 1.ppt." We are to copy the notes from the slides numbered one through sixteen, except the "homework slide," into our notebook.

9:45 a.m. End of class.

9:50 a.m. I proceed up to my office and open my journal to add a reflection on my experience in this first class. Here's what I write:

Reflection on my 2/7/11 class

- I need to do something about the glare on the board—I couldn't see the spelling of the words for ATP.
- I did not seem to be a distraction in the class. This is an important point in my research. The students know me well—most had me as their math teacher and everyone sees me around the building.
- This class seems like it's going to be a lot of work. Today, there was a lot of non-stop activity jammed into 75 minutes—I wonder what I've gotten myself into?
- I'm going to have to figure out how I can take notes in class for the required chemistry notebook and simultaneously take field notes in my research journal.

4:00 p.m. I started my homework. But I've had several interruptions so I've only completed eight of the sixteen required slides. I'll have to finish my homework later at home because I have a Venture Crew meeting in the cafeteria at 6:30 p.m. (I am an Associate Advisor for the club).

After the meeting, I made the thirty-minute trip home. My mind is on all of the activities we performed in Mrs. Waters' class today. Is every class going to be like this? Who will win the bet, Skeldon or Pilkington? And, am I going to have a bad experience working with chemicals in Mrs. Waters' labs?

After all, I've had problems in the past . . . like the time a few years ago when my daughter, Brittany, reminded me that she was late for a soccer game at the same time as I was in the middle of pouring chlorine into the backyard pool filter. Since I was rushing, I skipped the "measure accurately" step, put too much of the chemical in the automatic chlorinator and couldn't screw the cover on fast enough. I heard a sudden gurgling noise and turned my head just in time before the filter's PVC piping exploded sending a twenty-foot high stream of greenish-yellow chlorine gas into the atmosphere. Three fire trucks, two police cars, and an ambulance arrived within ten minutes. After listening to the Fire Chief lecturing me on the dangers of handling chlorine, I was taken to the hospital by ambulance because I had inhaled the gas. I remember all of the more-curious-than-concerned faces of the dozens of neighbors lined up along the street as the ambulance pulled out of the neighborhood. All I could think of was those national news stories of train derailments with toxic materials that require evacuating entire neighborhoods for days—would this be the case here?

I eventually recovered, and five days later, was able to repair the hole in the filter, but not the apparent hole in my head! Needless to say, I subsequently developed a fear of handling chemicals. It still haunts me today—I'm even contemplating cutting chemistry class on lab day.

One advantage of reflection, is that it can make the thirty-mile drive home seem shorter.

8:30 p.m. As I arrive at home, I continue my homework. I found a three-ring binder in my downstairs office and made three tabbed sections for the Chemical of the Day, Problem of the Day, and Class Notes. But I'll have to ask Mrs. Waters for clarification on whether a three-ring binder with tabbed sections can be used in place of separate notebooks specified in the syllabus. What other sections will be required? I googled some chemistry related images for the front cover. Unfortunately, I didn't finish copying the notes from the remaining eight slides—well there's always tomorrow.

Tuesday, February 8, 2011 (Day 3), 6:30 a.m. I arrived at school early today so I can finish my chemistry homework. I am able to complete the remaining slides, but I have a few questions for Mrs. Waters.

1:10 p.m. I got to class early today hoping to change my seat so as to avoid having to deal with the glare on the board. I chose the seat exactly opposite from where I sat yesterday. It was the closest one to the board, but on the opposite leg of the "U" shaped table configuration (see ch. 1, fig. 1.2). Unfortunately, it was where Sima had sat the first day—recall that I had mentioned earlier how territorial we are. Sima said she didn't mind sitting elsewhere. I was now sitting next to Siraj, Sima's cousin. This eliminated the problem of the glare.

Before class officially starts, most of us have already begun copying the Chemical of the Day and the Problem of the Day from the board into our notebooks.

1:15 p.m. We start class by working on the Chemical of the Day. Mrs. Waters comes around to check our notebooks. I confirmed with her that we can use a three-ring binder with dividers rather than three separate notebooks. The Problem of the Day involves writing a response to Louis Pasteur's famous quote "In the fields of observation chance favors only the prepared mind." She has written this on the board.

1:50 p.m. We are taking a pre-test for chemistry on the Study Island. It is a 20-question paper-and-pencil test. Everyone has to complete the same set of questions. The purpose is to establish a baseline for our knowledge of chemistry prior to our taking Mrs. Waters' class. At the end of the course, we will take a very similar test, but its purpose will be to assess the growth of our knowledge. This data will be used as part of an evaluation of Mrs. Waters' effectiveness as an educator. She had volunteered to participate in a pilot program for Rhode Island's new Educator Performance and Support System (EPSS), commonly referred to as the *Rhode Island Model*.[3]

After the pre-test, we go over slide nos. 1-16. These were assigned as homework last night and were due today. Mrs. Waters is projecting the slides and happens to call on me while I'm busy recording my observations and being only halfway attentive.

"John, is H_2SO_4 (sulfuric acid) inorganic or organic?"

I am busy writing in my journal—half listening and blurt out "it's inorganic." This, of course is the wrong answer and I think I hear Ruby [pseudonym], who is seated on the other side of the "U," calling me stupid! I'm actually happy to hear this because it's an indicator that she is accepting me as a fellow student. Unfortunately, this student interaction would be considered "unsatisfactory" using Danielson's rubric for "Creating an Environment of Respect and Rapport" (Danielson 1966).

The class was nearly over and Mrs. Waters gave an extra credit problem: to research the current status of the Large Hadron Collider in Switzerland.

2:30 p.m. End of class.

Wednesday, February 9, 2011 (Day 4), 11:15 a.m. Class began with a short review of some of the concepts that we had been learning. Mrs. Waters calls on me again. Her question was similar to the one I had gotten wrong yesterday, however, today I was paying attention and got it right.

Wednesdays are reserved for assignments on Study Island. We have a mobile cart with twenty laptops. It can be signed out for use in any classroom so that we don't have to travel upstairs to the frequently booked computer labs.

Mrs. Waters is requiring us to create and email her *error documents* or *mistakes documents*. They should contain up to three questions that we have gotten wrong on Study Island. The correct solutions should also be included.

I went to my office and grabbed my laptop computer (Mrs. Waters signed my agenda), then came back to class and downloaded JING.[4] This is a great screen-capture program that is available free of charge. I had to setup a username and password. During our webinar on Friday, February 6 (Day 1), when Carolyn Taylor, Mrs. Waters, Ms. Perkins, and myself were receiving professional development direction on the science component of Study Island, we discussed Mrs. Waters' requirement of having students collect the questions they had gotten wrong, together with the correct solutions to them. Ms. Perkins had suggested that this may be a perfect application for JING—a screen capture tool that she had been using in her math classes for the same type of application.

When students get a wrong answer on the screen, they call up the explanation for the correct answer (via a pop-up window) on the screen and use JING to size and capture the desired portion of the screen. They then copy and paste it into a "My Mistakes" template. After all the mistakes from a given lesson have been captured on the template, students forward the template (saved as Block1-Last Name-First Initial-Lesson Title) to Mrs. Waters via email.

My thought was to compile all of the mistakes into a spreadsheet and look for patterns where content area was concerned, things that may require remediation or special preparation for the State science test in May.

After downloading JING today, I was able to read the lesson on "Scientific Inquiry" and answer the first three questions. I got the first two correct but not the third. I used JING and saved my first mistake to the template. However, lots of intermittent problems with the internet have made it impossible to get as much done as I would have liked today.

1:05 p.m. End of class.

Later that night, I completed the lesson and got two more questions wrong! I noted that two out of the three pertained to the word "control." To me, the explanation that was provided for the correct answer was confusing and seemingly contradictory. Also, I thought the reading level might be high for many of our students.

Thursday, February 10, 2011 (Day 5), 9:50 a.m. Mrs. Waters is out of school today with a sick child. Mrs. Owsik (the health teacher) is covering her classes and I hear she is planning some "sex education" topic. I don't think it's appropriate for me to be in class today. That leaves Hector by himself with 16 females in Mrs. Waters' Chemistry Block 1 class.

Fortunately, both Mrs. Pagliaro (the business teacher) and Miss Wall (the humanities teacher) have asked me to do a SAT Math Prep lesson in both of their Senior Prep classes. Miss Wall's class is during Block 1—perfect, so now I have an excuse for not being in chemistry today!

Later that evening, I worked on my chemistry presentation—I found information on a new life form that NASA just discovered. Also, I completed the Study Island lesson on "Scientific Inquiry." At the end of the online lesson, I had earned a grade of "80" and emailed Mrs. Waters my "Mistakes Document" for "Project 1-Assignment 1."

Mrs. Waters has already set up the order of assignments on the Study Island. So, when we log in with our passwords, we see the next assignments that need to be completed. For me, the next thing that needed to be done was a Biology pre-test. I found it difficult considering there was no pre-lesson and I had not taken a biology course in approximately thirty-five years. I ended up with a grade of "66.7." My fellow students had just completed Biology last year so it would be interesting to see what their scores were.

I'm making a note in my journal that I think we should look at the lessons ahead of time and prepare vocabulary words that need to be reviewed before actually doing the assignments.

Friday, February 11, 2011 (Day 6), 5:00 a.m. I got up early this morning and spent an hour finishing my presentation for class today.

8:00 a.m. Class begins with DEAR—I'm spending the thirty minutes updating my journal. Meanwhile, Mrs. Waters is writing lots of notes on the board for Chemical of the Day, Problem of the Day, homework for Monday, and the agenda for today:

H.W. for Monday: slides #16 – #30 (skip #27)
1) What will we be doing? 2/11/11
2) Chemical and Problem of the Day
3) Homework presentations
4) Scientific inquiry activity
5) Science misconceptions activity
Review Study Island problems
–Objectives: Students will be able to
–collect and organize info
–evaluate and make inferences
–coordinate new w/old info and modify when needed
–write a defense of their inference
–evaluate and critique inferences presented by others

Today's Problem of the Day

Mrs. Waters has written:

> Devise a way to separate sand from a mixture of charcoal, sand, sugar, and water. You *may* show your thinking in the form of a flow chart.

Mrs. Waters tells us that on the State science exam, the word "may" should be interpreted as "you should" since the scoring rubric often gives more points to those who do it. So, here we need to assume that we must create a flow chart.

After a few minutes, Mrs. Waters asks if anyone has a method flow charted to share. No one answers; so, following my Rule no.1 ("I'll never volunteer first unless no one else steps forward"), I volunteer to go to the board and I draw my flow chart (see fig. 2.3). Everyone agrees that it seems like a reasonable solution and that it would be interesting to see if it works. Mrs. Waters says, time permitting, we may be able to set up a lab experiment before the end of the semester.

We proceed to go over the presentations pertaining to the "homework slide." The assignment on the PowerPoint (PPT) slide was to research and

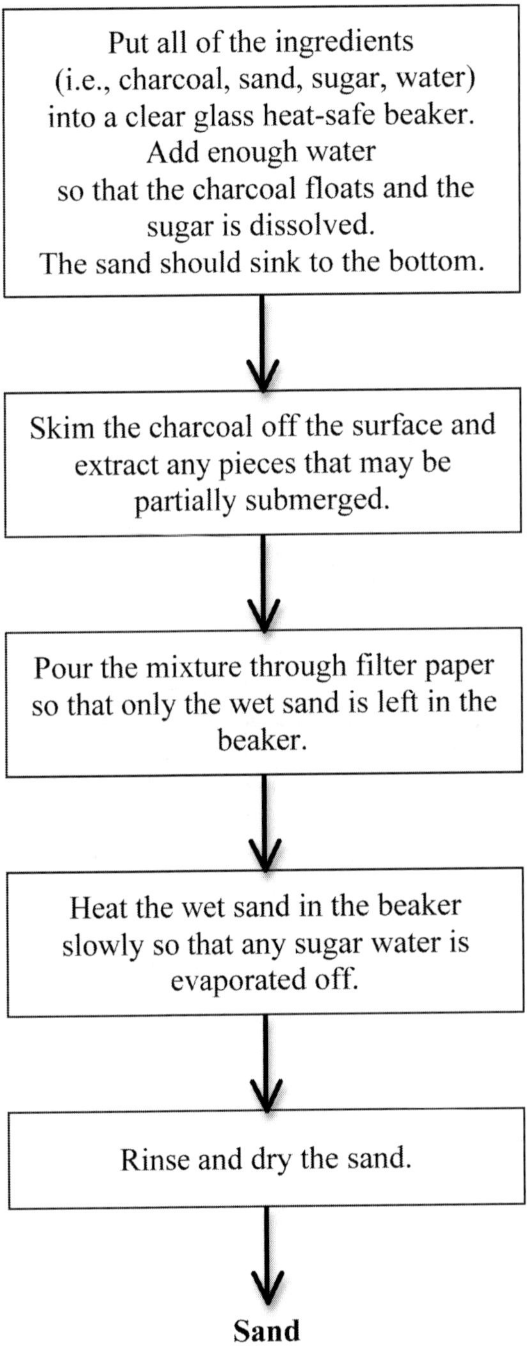

Figure 2.3. A flow chart for separating sand.

prepare a short report to the class on a topic related to chemistry. Mine was "NASA researchers discover a new kind of life." [5]

While listening to other presentations, I add the following notes to my journal:

- An interesting observation—although a few students haven't done their homework, most have and they are all doing quality work—they understand and present their topics well.
- Before class started, one female student (recently transferred to Beacon) stated that Mrs. Waters was the most beautiful science teacher she's had—they were all hags at her last school.
- Sima (who now sits next to me each day) tells me she doesn't think the lessons on Study Island provide sufficient information to answer all of the questions. I agree.

9:45 a.m. End of class.

Monday, February 14, 2011 (Day 7), 8:00 a.m.–8:30 a.m. It's DEAR. To save time, I've decided to copy the Chemical of the Day and Problem of the Day into my notebook. The Chemical of the Day is *benzene*. Mrs. Waters has written several characteristics of benzene on the board. One interesting fact is that it was used in aftershaves in the 19th and 20th century. For homework, she asks us to look up information on why benzene was used in aftershaves. I suggested that it was because of its sweet smell that I derived from one of the characteristics Mrs. Waters had written on the board.

8:30 a.m. The Problem of the Day is projected on the front board from a PPT slide on the term "control." This was a Study Island problem that many students, including myself, got wrong.

Mrs. Waters comes around to check the homework assignment—Whoops! I must have forgotten to write down part of the homework that was due today. It required that we look up the definition of greenhouse gases. So, I have completed only the PPT slide notes. Mrs. Waters makes a notation in her grade book to the effect that my homework was incomplete.

She continues to move around the "U" checking other students' homework and stops to praise Brittany's because she cut out a printed copy of the separate slide and pasted it in her notebook.

I add the following notes to my journal:

1. We need to add into our notebooks a section for vocabulary words.
2. I need to add a page to my chemistry binder for logging homework so I won't forget an assignment in the future.

Next, we did a great group scientific inquiry activity:

It involves making scientific inferences and drawing conclusions based on information at hand. And, so it also leads the students to discover that, when additional data come available, existing scientific theories may have to be modified. There are several variations of this activity, but Mrs. Waters has chosen one entitled *Scientific Method* by Brenda L. Dempsey (George Washington High School; Denver, Colorado 1998). The materials required for it include sixteen personal checks that contain a date, pay to, a dollar amount, memo, and a name and address of the checking account owner. This will be the data that is used to make inferences that enable one to piece together a story as to what is going on.

Dempsey provides the following instructions for this open-ended activity: In Round One, each group is given an envelope filled with five of the sixteen checks. They have been randomly selected so no two groups receive the same combination. Each group of students is asked to make five inferences supported by the data from the checks. In Round Two, four more checks are given to each group. Students are asked to re-evaluate their inferences based on the new data, and then to retain or modify them, or create new inferences if necessary. In Round Three, four more checks are revealed and again students have the same options. With thirteen of the sixteen checks in hand, each group of students must now create a *scientific theory* based on their conclusions as these have been drawn from their evidence-based inferences.

For this activity, Mrs. Waters assigns each of us to a group. I'm working with Katie and Hector. Our group was able to finish Round One and had developed our five inferences, but class was about to end. Mrs. Waters said we would finish the activity tomorrow.

9:45 a.m. End of class.

Tuesday, February 15, 2011 (Day 8). Before going to class, I looked up *greenhouse gasses* on the internet (the homework that was due yesterday—I can hand it in late but I'm losing points). I also researched the "Why benzene in aftershave?" homework that is due today.

1:15 p.m. The Problem of the Day concerns a question that everyone got wrong on Study Island. There is no Chemical for the Day today. Mrs. Waters is checking our homework. I have my "Why benzene in aftershave?" assignment done. And, I have my greenhouse gas homework assignment, the one I forgot last class. She confirms that she will give me partial credit.

She announces "From now on, Fridays will be quiz day. Everything we've covered is fair game."

We finished working on the Scientific Inquiry sixteen-envelope activity that we started yesterday. Mrs. Waters has each of the groups report on their conclusions. Every group has a different scenario, but there is general agreement on the fact that Brittany's group has the most creative story.

Wednesday, February 16, 2011 (Day 9). Before class begins, I meet with Mrs. Waters. I explain to her that I have been developing the "Guiding Principles for Homework" for Beacon (Butler 2011d, 2011e). In my research of the pertinent literature, I have come across some recommendations that students should have a voice in some of the assignments to be given. They should be given opportunities to create their own—this I refer to as *flexible assignments*. The hypothesis at work here is that students are more apt to be engaged in their assignment if they have ownership of it. I suggest to Mrs. Waters that she try giving a flexible homework assignment related to chemistry to test their engagement in the class (to be discussed in greater detail in ch. 3, "The Aesthetics of Learning").

11:15 a.m. Today being Wednesday, we will be working on the Study Island. However, Mrs. Waters begins by assigning the homework that will be due on Friday. She states simply: "You are to design your own homework assignment. It must be related to chemistry and you must be ready to present it on Friday."

I begin working on the next assignment on the Study Island. It is called *Lab Tools and Safety*. However, there are lots of problems with the internet today. Many of us can't log on and others are getting booted off. Eventually I am able to log on. I begin by studying the lesson and then take the test associated with it. However, I notice that there are questions on microscope parts and the three-beam balance scale. This is information that was not covered in the online lesson. Study Island is good but not perfect. Siraj was sitting near me today and tells me that when she gets stuck on a question, she uses Google, the widely used internet search engine, to find the solution or to get information leading to one. In other words, while working on Study Island it is possible to simultaneously use Google to get answers.

My concern with this revelation would be with the Study Island's pre-tests. The pre-test is always the first assignment to be completed when one is starting work on a Study Island topic. It is used for establishing a baseline of student knowledge. Students' use of the internet in this way could distort that baseline and make it harder to accurately measure their intellectual growth. My feeling is that any pre-test of this nature should be of the pencil-and-paper sort, or given on a computer in such a way that internet access is prohibited. However, for the Study Island lesson that is provided, I would recommend students be allowed access to the internet as a way of supplementing their research of unfamiliar material; they might also use it to locate information not contained in the Study Island lesson.

12:40 p.m. End of class—time for lunch!

Thursday, February 17, 2011 (Day 10). Both Mrs. Waters and I were late getting to class because we were being interviewed by a team from

RIDE (Rhode Island Department of Education). Beacon is one of three Rhode Island schools that is piloting Rhode Island's new educator evaluation system and Mrs. Waters has volunteered to be one of the teachers to be evaluated.

She got back to class before me. When I walk in, I have missed the Chemical of the Day. However, I am able to participate in the Problem of the Day. Next, we go over some of Mrs. Waters' PPT slide notes followed by more discussion on greenhouse gasses.

She has written on the board:

What will we be doing? 2/17/11

1) Chemical and Problem of the Day
2) PPT notes
3) Scientific Method
4) Discuss green house gasses
5) Review Study Island problems

Objectives: Students will be able to

- examine a real life investigation and determine what the hypothesis was and how it was tested.
- provide evidence as to whether or not the hypothesis should be supported
- distinguish between states of matter

11:10 a.m. End of class.

2:30 p.m. Mrs. Waters tells me there will be no class tomorrow because all of the juniors (i.e., eleventh-graders) are taking the ASVAB (Armed Services Vocational Aptitude Battery) between 8:00 a.m. and 11:00 a.m. We give juniors this test every year. It was developed by the Department of Defense as an entrance exam to the military and provides prospective recruits with a list of their potential occupational specialties and the signing bonuses they might qualify for. Critical areas tested are arithmetic, reasoning, vocabulary, paragraph comprehension and mathematical knowledge. We use the test as a diagnostic tool for identifying those areas in which the student may need improvement. It also identifies their strengths—all of which are helpful in pointing them to possible career paths. It takes about two-and-a-half to three hours. So, when Mrs. Waters tells me there is no class, I immediately think I don't have to work on my homework for tomorrow (Friday)—the last day before the February vacation week. That means I will have over a week to prepare my *flexible* homework presentation and two other homework assignments: Is FeO_3 iron trioxide? And, why is corn not properly digested by the human system?

What have I learned so far? It's only been ten school days (i.e., two school weeks), but certain patterns and routines have been established and are being accepted as part of the culture of Mrs. Waters' classroom. Here are the items I have included in my journal:

- Mrs. Waters has posted her agenda of class activities and objectives on the right-hand side of the whiteboard every day before we have entered. We are already in the habit of checking the whiteboard upon entering the class to see what we are doing that day.
- There has been a short activity at the beginning of each class. Upon entering class, we expect to work on a Chemical of the Day and/or a Problem of the Day. Thus far on a few days, just before class has officially started, I've noticed some students passing out blank notebook paper to others. They start working on the problems posted on the board without Mrs. Waters' involvement.
- The tables are arranged in a "U" pattern. Although Mrs. Waters has not assigned seating, students have consistently sat in the same general location, if not precisely the same seat each day. And, there are definitely patterns to be noticed: the same students choosing to sit next to each other every day. Also, they seem to respect each other where claimed territory is concerned.
- Mrs. Waters has administered a pre-test to establish a baseline of student knowledge. It is from this that she will be able to measure the growth of their knowledge. She has also used a questionnaire to identify various student interests—areas she might want to tap into.
- With all of the group activities undertaken thus far, Mrs. Waters has assigned the members of each group and has varied roles assigned to each. We have not been allowed to select study partners. I suspect Mrs. Waters has a strategy for determining group membership; however, it would be premature for me to state what it might be based on.
- Every student must have their *Student Agenda* signed by Mrs. Waters when they need to leave the class. However, she has not required us to sign back in upon our return.
- There is very little downtime in the class. It's bell-to-bell teaching and learning every day. There are multiple activities in each class with very short transition times between them.
- Mrs. Waters has posted assignments and resources on her Digication website. It is accessible 24/7.
- She promotes good note taking and notebook organization, and uses a rubric for grading the notebook. It draws on five categories, each worth twenty points for a total of 100 possible points. These categories are: *Required Items Present* (notes, handouts, in-class activities); *Neatness*; *Organization*; *Problems of the Day*; and, *Chemical of the Day*.
- Mrs. Waters keeps her PowerSchool online grade book current.

- She requires that the course syllabus be signed by parents and returned. Her grades and assignments are posted online and so accessible 24/7 by parents. All of this, taken together, provides clear evidence that she wishes to keep parents informed.
- It appears that Mrs. Waters will be providing several extra credit opportunities throughout the semester. We have already had a few in just the first ten days.
- A flexible homework task was assigned. I had suggested this type of an assignment and Mrs. Waters was willing to try it by way of experiment. The results are discussed in chapter 3 "The Aesthetics of Learning."
- There have been no significant behavioral problems in this class. I believe this is mainly due to the fact that Mrs. Waters has clearly established a well-structured learning environment within the first two weeks of the semester. She is consistent in her expectations, and activities run from bell-to-bell with little or no transition time. Most importantly, her lessons are engaging.
- We know that Wednesdays are reserved for Study Island, Thursdays will be lab days, and we will have a quiz on Fridays. There is structure and consistency in Mrs. Waters' instructional week.
- Homework is assigned each night, and Mrs. Waters checks it the following day.

Thus far, I am impressed with the serious attitude toward learning that my fellow students have displayed in Mrs. Waters Chemistry Block 1 class. I propose that, in addition to the expectations that she has established and engaging activities she involves them in, this may also be due to the students' perception of Mrs. Waters in these first ten days as a caring teacher with a passion for teaching chemistry.

Although it is still early in the process, these observations of her patterns and class routines are helping me to build a framework for assessing student engagement and teacher best practices. The latter is the subject of chapter 8, "Mrs. Waters' Best Classroom Practices."

NOTES

1. This is one of the several variations of the dihydrogen monoxide hoax that has been widely circulated since the early 1990s. The original author of this particular version of it is unknown. The worksheet was created by Mrs. Waters.

2. Answers to the *Ten Science Misconceptions* activity: Q1 (F); Q2 (F); Q3 (T); Q4 (F); Q5 (F); Q6 (F); Q7 (F); Q8 (F); Q9 (F); and, Q10 (T). The original author of the individual questions is unknown. The worksheet was created by Mrs. Waters.

3. *The Rhode Island Model* is the new educator evaluation system that RIDE (Rhode Island Department of Education) began gradually implementing in the 2011-2012 school year. Beacon Charter High School was one of a few Rhode Island Schools that participated in the program's pilot in the spring of 2011. The Rhode Island Model calls for annual evaluations, with a focus on educator-evaluator collaboration and feedback which can fuel professional growth; it also requires that the instructor have specific goals and objectives in place so as to be able to measure student progress. To determine overall educator effectiveness, the Rhode Island Model takes into account three central components: Professional Practice, Professional Responsibilities, and Student Learning. For some content areas and grade levels (e.g., ELA and mathematics), teacher evaluations are linked to student performance on the state assessments (RIDE 2011b).

4. JING, a product of TechSmith (http://www.techsmith.com/jing.html) is a screen capture program that was used in Mrs. Waters' class in conjunction with Study Island. The program was a free download and provided a very simple way of highlighting any part, or all, of the screen to copy and save images to file.

5. *NASA researchers discover a new life*. At a news conference on Thursday, December 2, 2010, NASA researchers announced the discovery of a new kind of life. Up until this time, it has been assumed that all life on Earth was based on six building blocks—carbon, hydrogen, nitrogen, oxygen, phosphorus, and sulfur. This was therefore shocking indeed! Now, however, microorganism has been discovered in the Mono Lake region of California that replaces phosphorus with arsenic. The new microbe has been named GFAJ-1 (Lendino 2010).

However, in her December 8, 2010 *Newsweek* article "Arsenic and New Life—or Not: Claims of a new life form get whacked," Sharon Begley cited several past news conferences that had announced major discoveries or claims which were then later found to be based on flawed research. These included *cold fusion* in 1989, the *Martian rock with signs of life* in 1996, and recent claims about the genetic basis for longevity. Begley provides several examples of explanations as to why there may be arsenic in the Mono Lake bacteria (Begley 2010).

Chapter Three

The Aesthetics of Learning

> Emphasis on the aesthetic dimension can also link the sciences with humanities, with philosophy, history, and the arts. In particular, it can make science more accessible to artists.
>
> —*Maura Flannery, Using Science's Aesthetic Dimension in Teaching Science*

LOOKING BACK

Beacon Charter High School for the Arts, located in Woonsocket, Rhode Island had its five-year charter approved by the Rhode Island Board of Regents on July 23, 2003. The school initially opened its doors in September of that same year. The first two years of operation were problematic. A financial shortfall had been created by a lower than anticipated enrollment. Subsequently, the school found itself in the position (early April, 2005) of having to agree to surrender its charter to the Commissioner of Education at the end of the school year. In exchange for this, $280,000 in state funds would be provided to allow the school to meet its operational obligations to the end of the 2004–2005 school year. Consequently, the charter was revoked in June 2005, well within the five-year contract period, by the Rhode Island Department of Education. They cited mismanagement of finances and poor administrative leadership. Thus, Beacon Charter High School became the first of the eleven charter schools operating in Rhode Island to close.

Several hearings were subsequently held (allowed by Rhode Island Charter School Law) concerning this. Since he most likely did not want the blemish of a charter school having been closed on his watch, Governor Carcieri, a proponent of these schools, helped get a reprieve from the Board of Regents for Elementary and Secondary Education. It allowed for the reopening of the school for the 2005–2006 school year (Nadeau 2005).

An amended charter was then approved which included the restructuring of fiscal oversight and co-sponsorship by Johnson and Wales University. They provided an academic and administrative partnership. The original sponsor was the Northern Rhode Island Council of the Arts. Also, the amended charter limited the school's enrollment to 115 students for 2005–2006, 140 for 2006–2007, and 180 for 2007–2008. The founder and President of the Rhode Island League of Charter Schools, Robert Pilkington, was hired as Principal.

New teachers were hired as long-term substitutes at substantially reduced salaries. They were given no health benefits for the first 90 school days in order to minimize payroll obligations. Several changes were made to streamline costs, e.g., eliminating a maintenance position, changing payroll services, closing an offsite business office, eliminating expensive copy machine contracts, etc. With newly restructured governance, a workable budget in place, and a qualified staff on board, the school could now focus on academics with virtually a clean slate.

Monday, September 12, 2005. This was an exciting day for me at Beacon Charter High School. The school reopened its doors to 115 students and a much-reduced staff. I was one of the new *reduced-salary no-benefits* teachers hired. I would be the school's grades 9–12 math teacher. In attendance were Rhode Island's Governor Carcieri, Irving Schneider, President of Johnson & Wales University, and several other dignitaries. Film crews from local television stations and reporters from the local newspapers (Nadeau 2005) were also present. Even the Governor's podium was brought in. This high-profiled event was symbolic of the Governor's support for charter schools, but also it was an indicator of the political scrutiny and enhanced accountability that Beacon would have to endure. But I remember this day for a different reason . . .

BOY UNDER THE COAT

In one of my math classes that first day, there was a young man sitting in the back of the class with his coat over his head. He did not respond to any directions from me, just sat motionless—the school's first *living statue*.[1] The other students called him *Jesus*. Later, when he left the classroom and exposed his shoulder-length dark hair I would make the connection. At the end of this first day, Principal Pilkington requested that the faculty meet for a debriefing session. He asked if there were any questions or issues this first day. I mentioned the *boy-under-the-coat*. One of the teachers who had been part of the original faculty, spoke up. "Oh that's Jesus, he's just waiting to turn eighteen so he can sign himself out of school. He put his coat over his head all last year and never did any work. His eighteenth birthday is in a few weeks—I'm thinking he's gone."

It was a very awkward situation. Over the next few weeks, each day, I would try to strike up a conversation with the boy-under-the-coat. I passed out classroom assignments to him only to retrieve blank sheets at the end of class—not even a name. Although we had very few conversations, I remember him as being very polite and well spoken. He just refused to do any work in any of his classes. He didn't seem to have a connection with any adults in the building nor did I see him associating with any other students. Sure enough, a few weeks later, on the very day of his eighteenth birthday, the boy-under-the-coat signed himself out of Beacon. Our paths would cross a few years later when he attended one of our school events. I asked him what he was doing and he told he was working for a large shipping company loading trucks. I have never forgotten him. He continues to haunt my memory—for me, the poster-child of those who fall through the cracks in education, and all the more disturbing when one considers that it happened under the auspices of the *No Child Left Behind Act of 2001* (NCLB) (107th Congress 2002).

IN SEARCH OF ANSWERS

Perhaps what bothered me the most about the boy-under-the-coat event, was that this was not my first observation of student disengagement from learning. In chapter 6, "The State Science Exam" and chapter 8, "Mrs. Waters' Best Classroom Practices," I discuss two other significant events that have been the impetus behind my continuing research into student attitudes—more specifically, student engagement (e.g., my presence in Mrs. Waters' chemistry class).

I had been accepted into the Joint PhD Program in Education at the University of Rhode Island and Rhode Island College on April 11, 2003; so, the event on the day of Beacon's reopening in the Fall of 2005 coincided with work on developing ideas for a dissertation proposal. Since I was a math teacher in an urban school, I wanted to focus on a problem related to mathematics education of *at-risk*[2] populations.

At the beginning of the first semester in Fall 2006, I asked forty-two of my Algebra 1 and Algebra 2 students to write an essay about their past experiences in math classes. The following are brief excerpts from four of them:

> "Math has always been a tough subject to me. In the beginning of every school year I wonder and dread about my new math class" (11th grade Algebra 2 student).

> "I always feel like I fail at math. I work really hard to get the 'A' that I want and need to get my college scholarships. But I feel miserable and unsuccessful in mathematics all the time. It makes me think less of myself because I know math is everywhere" (11th grade Algebra 2 student).

> "My dad was never really good at math, so I could never ask him to help me. I put one of my study guides in front of him, and he said that it made his brain hurt just by looking at it" (9th grade Algebra 1 student).

> "All of my classes were pretty good, except for math. The teacher was one of the worst teachers at the school. She was mean and made math horrible for me. That was the year I hated school. I didn't want to go and I didn't do any of my work" (9th grade Algebra 1 student).

Each year I collected these *My Math Experience* essays, I was confronted with similar disturbing content—a significant number of students were entering Beacon in the ninth-grade with bad experiences in their backgrounds and/or with negative attitudes towards math. The word "hate" often appeared in these student reflections. I feared that these negative attitudes toward math could evolve into disengagement with school in general and eventually result in some of these students dropping out. Thus, I began an extensive review of the research literature on anything I could find on student engagement. I wanted to understand why students shutdown and checkout.

REASONS WHY STUDENTS DROP OUT

During the months of August, September, and October of 2005, Civic Enterprises in association with Peter D. Hart Research Associates for the Bill and Melinda Gates Foundation, investigated the reasons why youths, ages 16–25 failed to complete high school. They analyzed this problem from the perspective of former students. This study involved a series of four focus groups and a survey of 467 ethnically and racially diverse young people (aged 16–25) who identified themselves as public high school dropouts. It was carried out in 25 different locations across the US including large cities, suburbs, and small towns, all of which had high dropout rates. The release of the March, 2006 comprehensive report entitled *The Silent Epidemic: Perspectives of High School Dropouts* (Bridgeland, DiIulio, and Morison 2006) lists the top five reasons why students leave school (see table 3.1). The authors state:

> There is no single reason why students drop out of high school. Respondents report different reasons: a lack of connection to the school environment; a perception that school is boring; feeling unmotivated; academic challenges; and

Table 3.1. Top Five Reasons Dropouts Identify for Leaving School. (adapted from Bridgeland, Dilulio, and Morison 2006)

	Response Percent	*Reason*
1.	47%	Classes were not interesting
2.	43%	Missed too many days and could not catch up
3.	42%	Spent time with people who were not interested in school
4.	38%	Had too much freedom and not enough rules in my life
5.	35%	Was failing in school

> the weight of real world events. But indications are strong that these barriers to graduation are not insurmountable (p. 4).

The study found that a majority of these dropouts had post-secondary career aspirations that required at least some college and had grades of C or better when they dropped out. The number one reason that students listed for dropping out was that they were bored with their classes—not motivated or inspired to work hard.

When respondents were asked what they thought schools could do to improve the chances of their staying in school, the answers usually related to classroom instruction—making what is learned in classes more relevant to their lives, having better teachers who keep classes interesting.

AN UNDERREPRESENTED RESEARCH TOPIC

Aware of the aforementioned research on dropouts, for my dissertation I was hoping to center on a question that might help math students like the boy-under-the–coat take an interest in and grasp the relevance of math in the context of their classroom experience. My aim was also to define a research area that was underrepresented in the literature. I found several, but one in particular, namely *aesthetics,* caught my attention.

> "A genuinely experimental approach to aesthetics as a whole has never been tried, and its possibilities of success or failure, good or evil, are therefore unknown" (Munroe 1956, 5).

> "We argue that aesthetic judgments, which have heretofore received little attention in the research literature, may be important factors in expert mathematical behavior. In particular, we argue that experts do monitor and evaluate their problem-solving activity and their monitoring and evaluation often have an aesthetic character" (Silver and Metzger 1989, 59).

> "Mathematicians have long claimed that the aesthetic plays a fundamental role in the development and appreciation of mathematical knowledge. To date, however, it has been unclear how the aesthetic might contribute to the teaching and learning of school mathematics" (Sinclair 2004, 261).

> "One approach that is still neglected is to stress the aesthetic dimensions in science. Inclusion of these dimensions has been, at best, a 'most muted' aim of new science education programs" (Flannery 1992, 1).

Studying student engagement in mathematical problem solving through the lens of aesthetics was just one of several perspectives I pursued in my doctoral research. In one study, I gave students a set of math problems to solve and asked them to choose which ones they liked or disliked along with explaining their reasoning. In another related study, I specifically chose a set of math problems that had multiple ways of reaching a solution such as using algebraic equations, constructing a table, graphing, or drawing a picture. I asked students to rank the provided solution method according to likeability. A statistically significant pattern revealed itself—the more students liked a problem, the higher was their problem-solving performance (Butler 2009).

Within the year after earning my PhD, as math department chair at Beacon, I was creating math word problems for a "problem of the day" to be administered at the beginning of all math classes at Beacon. But I decided to try a little experiment. I knew that these sorts of problems are not exactly the favorite activity for the typical high school math student. In fact, students often find them repulsive and don't even try to solve them—especially if there are a lot of words and no pictures (e.g., figures, diagrams, charts, tables). But what if I gave the students some ownership to the problem? I began creating problems that used actual student names in the particular math class. An amazing result—students were more engaged and more likely to solve the problem if their name appeared in the problem. I was getting calls from my math teachers telling me that their students "so-and-so" wanted to know if they could have their names in the next problem. For me, this was more evidence in support of the conclusion that aesthetics may play a key role in student engagement in problem solving. Admittedly, this fed a preconception when I embarked on my research in Mrs. Waters' Chemistry Block 1 class—but would there be opportunities to make similar observations in a science class as concerns aesthetics and student engagement?

AESTHETICS

In Mel Brook's 1981 film *History of the World: Part I*, a spoof of great moments in history, he introduces us to the birth of art. The scene takes place inside a cave, some two million years ago. A caveman-artist is painting a pre-historic animal on a cave wall. Several of his fellows are involved in the artist's work, smiling, nodding in affirmation, and grunting their praise as it evolves. When it has been completed, the caveman-artist steps back; and, with a wave of his arms, welcomes final responses to his creation. From the rear of the cave, another caveman approaches. Upon viewing the drawing, he scowls, pauses directly in front of the work, and then proceeds to urinate on it. The narrator is heard to say ". . . And of course, with the birth of the artist came the 'inevitable afterbirth'—the critic."

Unfortunately, the credits at the end of the film do not provide the exact time and place for this event. However, one need only view the exquisite 15,000 year-old paintings of the steppe bison in the Cave of Altamira in Spain to realize that perhaps Mel Brooks is not so far off—art appreciation began in a cave. For, at some point in ancient history, there had to be a *first* artist. And, unlike the "chicken and the egg" conundrum, there is no doubt that the artist and his art came first and the critic second.

With the critic(s), came the reality of *subjectivity*, for not all artwork would be urinated on by way of a response to it. Throughout history, man has contemplated the essential question "What is beauty?" Such contemplations are subsumed under the branch of philosophy called aesthetics. Until the early 1900s, aesthetics was defined as the *philosophy of beauty*. Its main task was to establish criteria that would correctly or more accurately define "the beautiful."

The long-held traditional definition of aesthetics has been: the branch of philosophy that deals with the nature and expression of beauty, as in the fine arts; and, of or concerning the appreciation of beauty.

However, "Artists, critics, and the general public are aware of the long failure of aestheticians to agree upon a definition of beauty or an objective criterion for estimating it" (Munroe 1956, 263). Over the ages, a continuum of thought has evolved between two extremes: At one extreme, the focus is on the object of art and the question addressed is: What characteristics or qualities does the object exhibit that make it attractive (e.g., beauty, elegance, order, clarity, symmetry, balance, proportion, economy, limitation, unity, and harmony). At the other extreme, the focus is on the perception and feelings of the observer toward the object (e.g., interest, surprise, unexpectedness, inevitability).

There have been attempts, albeit failed ones, to quantify the subjective world of the artist and critic. But by what instrumentation can we define beauty, elegance, or harmony of, e.g., a set of sculptures? How does one mechanically quantify *balance* between paintings of different subjects? Yet one finds in the literature many attempts to do just this—to provide a quantitative basis for aesthetic judgments. For example, in his *Mathematics of Aesthetics*

(1956, 2185–2197), George D. Birkhoff relates the complexity of an object (C), the harmony or order (O) of the object, and the aesthetic measure of the object (M) by the equation:

$$M = \frac{O}{C}$$

THE AESTHETIC EXPERIENCE

In 1912, Edward Bullough (1880-1934), English psychologist, philosopher of art, and Cambridge lecturer published a lengthy paper in the *British Journal of Psychology* entitled "Psychical Distance as a Factor in Art and an Aesthetic Principle." There he proposed a theory of art appreciation based on a concept he referred to as the *psychical distance*. As concerns the relation between an object of art (e.g., musical composition, painting, poem, play, sculpture) and an observer, he distinguished between spatial distance that could be measured in inches and feet, and a cognitive or psychical distance. Bullough proposed that an aesthetic experience was affected by two sets of conditions: those offered by the object and those realized by the observer. Further, not only do individuals differ from each other in the measure of psychical distance, but the same individual also differs in this measure with regard to different objects. Hence, Bullough concluded that psychical distance was bounded. Outside of this bounded region, no aesthetic experience is possible. These positions were labeled by him *psychically overdistanced* and *psychically underdistanced*.

In his 1992 book entitled *The Art of Mathematics*, author Jerry P. King dedicates his longest chapter (ch. 5) to the topic of aesthetics. He there includes his version of Bullough's psychical distance, respectfully referring to it as *aesthetic distance*. King represents Bullough's bounded region as doughnut shaped, formed by two concentric circles with the object at the center. He calls it the *Aesthetic Ring*. Anything that goes beyond the outer circle (i.e., aesthetically overdistanced) and within the inner circle (i.e., aesthetically underdistanced) produces aesthetic failure in keeping with Bullough's concept.

"According to Bullough, an observer too close to a work of art will be too practical or subjective toward it to appreciate it properly. If an observer is overdistanced, he will see the object as something cold and withdrawn and he will again fail to appreciate it as art" (King 1992, 200).

Bullough's concept received much attention. For example, Donald Sherburne, author of *A Whiteheadean Aesthetic,* states, "Edward Bullough's theory of the Psychical Distance has become a classic doctrine of aesthetic theory that must be taken into account by all aesthetic thinking." And, James L. Jarrett, author of *The Quest for Beauty*, writes of Bullough's ideas, "Perhaps no more influential idea has been introduced into modern aesthetics than that of psychical distance" (King 1992, 196).

In her book *Variations on a Blue Guitar: The Lincoln Center Institute Lectures on Aesthetic Education*, a series of lectures spanning 25 years for art teachers, author Maxine Greene cites John Dewey's view that the opposite of aesthetic is *anesthetic*. Greene states: "Anesthetics, for me, implies numbness, an emotional incapacity, and can immobilize, prevent people from questioning, from meeting the challenges of being in and naming and (perhaps) transforming the world" (Greene 2001, x). As I read Greene's definition of anesthetics and think of Bullough's psychical overdistancing, I recall that haunting image of the boy-under-the-coat. Greene defines aesthetics as "a term used to single out a particular field in philosophy, one concerned about perception, sensation, imagination, and how they relate to knowing, understanding, and feeling about the world." (Greene 2001, 5).

Up to this point, I have used the terms *art* and *object* interchangeably. Further, I have avoided defining art. This, however, poses a problem. Not only is it the case that over two thousand years of contemplation by some of the world's brightest minds has failed to produce agreement on what is beautiful, but disagreement is abounds on the question "What is art?" (e.g., fine art, commercial art, applied art); and there is further disagreement on "What is the purpose of art?"

Fortunately, this is a problem for the art world and one that need not contaminate the non-art world (e.g., mathematics, science). However, with regards to mathematics and science education, or any content area for that matter, I see a close relationship between aesthetics and student motivation, interests, engagement, effort, and different learning styles. I have witnessed too many examples of students applying "artistic thinking" in successful problem solving during my doctoral research not to acknowledge that aesthetics can play a major role in teaching and learning. One example, perhaps my favorite, is that of a taped interview of Jerry, a ninth grade Algebra 1 student, following a problem-solving task:

Jerry was one of several participants who had been given a set of non-routine mathematics problems to solve (Butler 2009). This particular problem, entitled "Blind Sided," involved a drawing showing three different views of a cube that had different colors on five of its sides—the sixth side not shown. The task was to state the color of the blind side—the one not shown in any of the three views. Jerry had submitted the correct answer, but had not shown any work. The purpose of the taped interview was to have Jerry explain his reasoning for getting his answer. He explained that he made a model. Upon further questioning he stated he used origami to solve the problem by making a 3D model. His strategy of using origami came as quite a surprise as the teacher proctoring the assessments did not observe him making a model. So, I proceeded to ask him to demonstrate his method. He asked me to get him a piece of copy paper.

For the next four minutes and 15 seconds, Jerry made a number of folds and creases, transforming the flat piece of paper into a flat, layered, somewhat diamond-shaped configuration. When finished with the folds and creases, Jerry explained that he had to blow into one of the corners to form the cube. He blew into the folded configuration and the cube took a shape, albeit not exactly a cube. He then rubbed the sides with his fingers into a perfect cube. Next, he labeled the sides with a letter representing each of the five colors and matched them to the sides visible in the three views. From his model, he was able to correctly determine the color of the blind side.

When asked how he decided to solve the problem using origami, he explained that when he was around ten years old, he read a book on origami and remembers making small bird-like things, boxes, and other "stuff." He continued, that as he read the Blind Sided problem and saw the views of the cube, it reminded him of the origami boxes he used to make.

Jerry also completed five other mathematical problems and was asked which one was his favorite and why. He responded: The Blind Sided Problem— "I got to do my favorite thing, which is...origami" (Butler 2009, 158). This, I claim is an excellent example of someone inside of King's Aesthetic Ring (i.e., Bullough's concept of psychical distance).

In this example, the "object" was a math word problem, but it could have been a math formula, a science experiment, the Periodic Table, a graph, a PowerPoint presentation, a reading assignment, a homework assignment, etc. It is possible for a student (i.e., the observer) to have an aesthetic experience when studying any of these—or not to, of course. And that experience can be either positive or negative. Knowing how students feel before, during, and after their encounter with the "object" can provide the teacher with data that can improve instruction and/or provide insight to needed intervention (e.g., differentiated/individualized work, remediation).

The two definitions that I take to be relevant to student learning are:

1. *Aesthetics* in learning may be generally defined as one's feelings towards the material to be learned and/or activity to be performed before, during, or after the assigned task.
2. *Aesthetic experience* is defined as the length of participation in an event involving an object (e.g., science experiment, math formula, reading assignment). The experience may be positive (i.e., pleasurable) and can consist of three successive phases. First, there must be a preliminary effort to give attention to the object. Secondly, there is a feeling of value that

rewards this effort (e.g., clarity, surprise, unexpectedness, inevitability); and third, there is closure—a realization that knowledge has been gained with respect to the object (see Birkhoff 1956, 2185). A negative aesthetic experience would differ from a positive one in that no reward would be forthcoming for the effort a student had expended and/or there would be no feeling that learning had taken place.

FLEXIBLE HOMEWORK AND PROJECTS

In addition to my duties as Math Department Chair, Data Analyst, and math teacher, I am also the Chair for the SIT (School Improvement Team) at Beacon. The SIT, which is comprised of teachers, administrators, parents, students, and community members, meets once a month throughout the school year. It focuses on improving the learning environment and identifies specific areas of need such as instructional focus, curricular initiatives, and staff development.

November 30, 2010 SIT Meeting. First on the agenda for today's 3:00 p.m. meeting was a discussion on homework led by Dr. Pilkington. Earlier in the morning, he had met with parents at his monthly "Breakfast with the Principal." Dr. Pilkington told us that several of the parents he met with would like it if more homework were required of students. He tells us "When parents ask their kids if they have homework, they often say they did it in school. Parents are seeing an inconsistency in the frequency and rigor of assignments between teachers."

Parents attending today's afternoon SIT meeting added that often their children do not understand the homework assignment sufficiently to be able to complete it; and, most of the time, they are unable to help. This causes stress in the home. Further, students say they find homework boring and fail to see its relevance. After a lengthy, but constructive discussion, the SIT agreed that the school needed to develop a set of guiding principles for a shared value system that would promote best practices and drive school-wide policies relative to this question. I was tasked with researching pertinent literature on the subject and asked to prepare a report to the SIT to be presented at a future meeting.

I was very interested in what the research literature would say. None of the comments that parents had made were new to me. Homework has long been a controversial topic in education: How often? How much? What kind? Can it be started or even completed in class? Should it be graded? And, should it be assigned at all? Admittedly, in ten years of teaching math, I don't recall any formal coursework or professional development seminars on this topic.

As with most teachers, my homework policies were my own—based on "gut feeling" and experience rather than being either research- or evidence–based.

For the next few months, I read extensively about homework in K–12 US education in scholarly journal articles and books. I found pertinent literature that spanned over 150 years beginning in the mid-nineteenth century in the first publically funded school systems. Two of the best historical accounts were from an article in the *American Journal of Education* (2000) entitled "The Lost Cause of Homework Reform"; there was also a 2004 *Theory into Practice* journal article entitled "Villain or Savior? The American Discourse on Homework, 1850-2003." Both articles were authored by Brian Gill and Steven Schlossman. In addition to these items, there were extensive contributions to this literature by Harris Cooper (Cooper 1989a, 1989b, 85–91; Cooper, Robinson, and Patall 2006, 1–62); and, Alfie Kohn's 2006 controversial book *The Homework Myth.* Taken together, all of these provide a fairly comprehensive overview of the continuum of attitudes toward homework as these have developed and changed over time. In fact, the breadth and depth of the literature on this was rather surprising. I learned quite a lot and my research evolved into a 35-page white paper entitled "Homework: Foundational Research for Beacon's Guiding Principles." I would eventually share it at a future SIT meeting. The culminating deliverable included in my paper was a list of eight guiding principles. This white paper is available on the school's website (Butler 2011e).

So, what does my discussion of research on homework have to do with Mrs. Waters' chemistry class? Well, I learned something very interesting from reading Gill and Schlossman—something that was worth pondering in the context of Mrs. Waters' class. I believe it can promote a positive aesthetic experience on the part of learners.

Gil and Schlossman state that just before and after World War II, as part of the progressive education movement, the central mission of homework reformers was to move it beyond repetitive paper and pencil exercises, memorization, and bookwork to more "student-centered" learning. The authors cite three key tenets of progressive education: (1) learning by doing; (2) educating the "whole child"; and, (3) child-centered, learner-driven education. I briefly paraphrase these authors' descriptions below:

Learning by Doing: Experiential Homework

This was an approach that was clearly in-step with the philosophy of John Dewey. These homework reformers believed that the classroom was the laboratory where concepts, and knowledge are learned and skills are acquired. The home should be a place where the child can apply these skills to everyday living. Involving the child in family shopping, in order to apply math skills learned at school, would be an example of experiential learning.

Homework for the "Whole Child"

The view of reformers in regards to this tenet was that the school was responsible for the education/growth of the "whole child" with respect to physical, social, emotional, mental, and spiritual development. Homework could include all of those experiences in the lives of children outside of the schoolroom that tended to foster desirable growth. Examples would include: the pursuit of individual interests through reading, listening to the radio, attending movies, engaging in conversation, developing appreciations or skills in music, or fine arts. There is really no limit to the range of interesting experiences a child could engage in. Students would choose their homework assignments under the guidance of teachers and parents.

Voluntary Homework: "Child-Centered" and Learner-Driven

This tenet centered on shifting the focus from the subject matter to the child. The homework reformers believed that homework should be learner-driven, independent, and voluntary. Optimistic proponents claimed that if students were partners in the planning of their school experiences, they would voluntarily add extra time at the end of their school day by doing homework whenever there was a need. The authors present one teacher's model in which, instead of a uniform assignment for everyone, his students volunteered to assume responsibilities that make distinctive contributions to their class.

Support for a move toward applying these progressive ideas came from reported "scientific" evidence that the traditional routine homework did not equate to increased student achievement. Gill and Schlossman (2000, 34) include the following quote from University of Alabama scholars, "There is no conclusive evidence from achievement test results or other achievement records to justify the persisting faith of many persons in the merits of routine homework (Swenson et al. 1955, 22)."

When the homework reformers shifted from subject matter to the learner, it naturally followed that assignments should be customized to suit the needs of individual students—a direction that fit neatly with the progressive educational theories of the 1920s and 1930s. Gill and Schlossman make the point that progressive educators understood that each child had different needs and abilities; and, in the child-centered school, they never expected all students to master the same material and/or at the same rate. An interesting and, I believe, little known fact, is that in the decade that followed World War II, (as reported in Gills and Schlossman) *individualization* was the most often repeated theme in the homework reform movement—"any plan giving uniform home assignments is a 'lock-step' system, as outmoded as the horse and buggy . . . Whatever homework is assigned should be based upon the children's interests and their ability to do the work (Merrick 1947, 26)."

It is generally accepted that World War II began in 1939, with the invasion of Poland by Germany, and ended in 1945—now some 67 years ago. Yet in 2012, I hear the terms "student-centered," educating for the "whole child," "differentiated instruction," "individualized learning plans," at just about every educational professional development workshop or conference I attend. Maybe those progressive educators were onto something decades ago! Having built upon their tenets and borrowing some language from Catherine Vatterott, author of *Becoming a Middle Level Teacher: Student focused teaching of Early Adolescents*; and, *Rethinking Homework: Best Practices that Support Diverse Needs* (Vatterott 2007), I am able to cite one of my eight guiding principles for homework as follows:

> *Whenever possible, homework should be personally relevant and customized to promote student ownership*: Students need to approach homework assignments with a positive attitude. Homework assignments should be designed to give students choices and opportunities to personalize their work—to give students more control over what they learn, how they learn it, and how they show that they have learned it (Vatterott 2007). Students are more apt to be engaged if the assignment is aesthetically pleasing (e.g., relevant, interesting, organized and neatly presented). The three main types of homework are:
>
> 1. Mandatory-Specified (MS): Homework must be done and per specific instructions.
> 2. Mandatory-Flexible (MF): Homework must be done but the student gets to create his own assignment relative to a general one supplied by the teacher.
> 3. Extra Credit (XC): voluntary, no penalty for not doing it.

I distinguish between the three types of homework as follows: Mandatory-Specific (MS) is the traditional type whereby the whole class is assigned generally the same homework and every student is expected to complete it. It may be differentiated in terms of rigor or adjusted for a student's learning style, but it is required of everyone and designed by the teacher with specific criteria. Mandatory-Flexible (MF) is also required of every student. However, the student has a voice in designing these assignments. Typically, the teacher will select a broad topic from the pertinent content being taught, but students will have the flexibility of creating their own assignment preferably linking their interests to the content to be learned. The third type of homework assignments, Extra Credit (XC), by those teachers who offer such an opportunity, is strictly voluntary. It is commonly used as a means of challenging students and/or providing them opportunities to raise their grades. There is never any penalty for any student not doing an XC assignment other than the potential widening of the achievement gap (e.g., content learned, course grade) between students who do XC and those that do not.

February 15, 2011 SIT Meeting, 3:15 p.m. This was our monthly meeting for the SIT (School Improvement Team). In attendance were: myself (SIT Chair), Dr. Pilkington (Principal), Mr. Skeldon (Academic Dean), Mrs. Hawkridge (Artistic Director), Ms. Russell (School-Based Coordinator), Ms. Fox (Reading Specialist), Ms. Perkins (Math Teacher), Mrs. Lareau (parent), and Caley Lareau (student). The main agenda item was my presentation to the team of the draft of my 35-page document on homework. It included a proposal for Beacon's eight "Guiding Principles for Homework." After review, the team clarified some of the language and requested that I prepare the document for posting online so that it could be shared with faculty and families.

On the drive home, after the SIT meeting ended, I kept thinking about the "flexible" homework idea. What if students were given the opportunity to create their own homework with virtually no teacher-defined constraints as to time-on-task, no specifics as to the deliverable (i.e., product), and no criteria for method to be used? Would they be more apt to be engaged in the learning process? Wouldn't this be a means of assessing student engagement as well as determining their interests? Knowing individual student interests, or being able to group students with similar interests could help refine the process of instruction. I definitely felt that treating "flexible homework" as the "object" could lead to some insights into how aesthetics might play a role in a student's learning.

February 16, 2011 (Day 9) Wednesday 7:45 a.m. I rushed down to Room 205 to catch Mrs. Waters before the start of school. This being a Wednesday, students and faculty will be heading over to the Stadium Theatre for our 8:00 a.m. town meeting. I wanted to suggest to her that she try making a "Mandatory-Flexible" homework assignment as a means of assessing student engagement. Finding her in her room, I made my suggestion and she said she liked the idea. However, I left not knowing whether she truly liked the idea or was just being polite.

11:15 a.m. It's the beginning of chemistry class and Mrs. Waters' announces that there is a homework assignment due on Friday: "You are to design your own homework assignment—it must be related to chemistry and you will present it to the class on Friday."

I was happy to find that she had not wasted any time on trying out this experiment. We are only nine days into the semester, just starting to get our feet wet with chemistry. So, it will be interesting to see what my fellow students choose for topics and how much effort and time will be put into a homework assignment that has no constraints or criteria other than having to be related to chemistry.

February 18, 2011 (Day 11) Friday, 7:40 a.m. It is the day before the start of our February vacation week. Mrs. Waters stops by my office before the start of school. She tells me she feels bad because the students in our Chemistry Block 1 class won't be able to present their flexible homework assignments today. All juniors are taking the ASVAB (Armed Services Vocational Aptitude Battery) down in the cafeteria this morning. She has already seen some students bringing in some baked items for their homework assignment. We agreed that we needed to find a way to get their efforts documented. Unfortunately, this was the last day before our week-long February vacation, and any baked items probably wouldn't survive a week. We had the main office make a school-wide announcement for all juniors in Mrs. Waters' Block 1 class to report to Room 205 at 11:10 a.m. following the ASVAB test (a time also coincided with the normal class passing time).

11:10 a.m. I am down in Room 205 with Mrs. Waters, and students are arriving with their flexible homework assignments:

Hector's Homework: Hector tells Mrs. Waters that he did an experiment at home and made a video. Unfortunately, he tells her that he didn't know how to upload the video but he has brought the camera to show the film and to see if someone can help him make a copy. Mrs. Waters watches the film and has a big smile on her face—she calls me over to view it. Apparently, Hector, whom I have always known as a very quiet, borderline shy student, has a passion and talent for explosives. He made a film of his creation of a HCl (hydrogen chloride) and Al (aluminum) bomb. On the film, you see a clear plastic bottle on the ground in an open space. Suddenly, the bottle fills with smoke and quickly expands ending in a massive explosion, a loud noise, and a plume of smoke. This supports those who argue that homework should be done at home and not in school! Mrs. Waters tells Hector to save the video and after vacation week she will figure out how to copy it.

Siraj's Homework: Siraj presents next. She tells the class that she was interested in investigating the effectiveness of detergents in removing stains. She came up with the idea for this topic because of a question posed by Mrs. Waters in class the other day. She had asked for examples of chemical reactions—when called upon—Siraj had provided an example of detergents when doing laundry. Today, she has brought in some bleach, water, a bottle of catsup, a plastic container, gloves, and a white cloth. She has set up her demonstration on a table in the front of the room near the whiteboard. Siraj dumps the catsup on the white cloth creating a very disturbing red stain. Then, with gloves on both hands, she dips the cloth into a solution of 50 percent bleach and 50 percent water. She had previously created it in the plastic container. After about fifteen seconds, she removes the cloth that is now perfectly clean. Siraj has also prepared a written report on her experiment and proceeds to read it to us. She then asks if there are any questions. One student asks if she thought that, had the catsup been allowed to sit on the cloth for a longer time, would the cleaning be as effective? Siraj says she does not think time is a factor. She tells us that she would like to do some future research in which the concentration of the bleach would be varied. This student-student discourse led to Mrs. Waters talking about some recent related vocabulary words we had been discussing in a previous chemistry class: "control" and "independent and dependent" variables. I asked Siraj how long she spent on her homework and she said it had taken about an hour.

Sarah's Homework: Sarah has brought in a large tray of cupcakes (see fig. 3.1). The tops have been frosted with the symbols of elements from the Periodic Table. Further, she has strategically frosted and/or arranged them on the tray to form familiar chemical compounds such as FeO_3 (iron oxide), H_2S (sulfuric acid), NaCl (sodium chloride or salt), CH_3 (methane), and NaOH (sodium hydroxide). Unfortunately, when no one was looking, a cupcake disappeared—and not as a result of "radioactive decay"! "John, is that chocolate frosting on your face?" Sarah explains that her older brother, a second-year pharmacy student in college, got involved last night when he found her frosting the symbols onto the cupcakes. He helped her make little white frosted dots on the various cupcakes to represent the electrons in the outer orbit. Mrs. Waters tells us that we will be discussing this when we cover something called "Bohr Diagrams" and "Lewis Dot Structures" later in the semester. I asked Sarah how many hours she spent on her homework assignment. She said it was about four.

Katie's Homework: Although this duplicated Sarah's assignment, Katie had independently chosen to bake a tray of cupcakes and frosted them with the symbols of various elements including Fe (iron), Zn (zinc), Ag (silver), and B (boron) (see fig. 3.2).

Sima's Homework: Sima has brought in an article on "mood rings." She tells us that she has always liked mood rings but never knew how they worked. She explains that the color of the mood ring is affected by the temperature of the body and that heat and cold are related to human feelings. Fluctuations in temperature affect the motion of the liquid crystal molecules inside the ring. The slightest change causes the molecules to become re-arranged. This causes a change in the wavelengths of light being absorbed or reflected by the liquid crystals and the result is a change in color. Dark blue indicates happiness, romantic, or passionate; blue points to calm or relaxation; blue-green indicates that a person is somewhat relaxed; green indicates normal or average levels of tension; amber indicates nervousness or anxiety; gray, great nervousness or anxiety; and, black, stress, tension, or feeling harried.

Manar's Homework: Manar has brought in an article on NASA's announcement that it has found a new life form that thrives on arsenic and actually uses the toxin in its DNA— remarkably, phosphorous has been replaced with arsenic. This new life form, a bacterium, was found living in the highly poisonous Mono Lake near Yosemite National Park in California. She explains that scientists believe that this discovery means that life can occur on other planets which have environments that, up to now, were thought to be inhospitable to life. Manar says she researched three to four articles before choosing this one (I had also chosen this same article a week ago and had a similar experience researching the question).

Ruby's Homework: Ruby chose to investigate something called the "Maillard" reaction in cooking. She explains that it is a chemical reaction that occurs between an amino acid and a reducing sugar, usually as a result of heat; it causes browning. Louis-Camille Maillard, the chemist for which the reaction is named, first described it in 1912. Browning is very important in the preparation or presentation of many types of food. Ruby chose this topic because she likes cooking and wanted to know why this occurs. She has also written a report on her topic.

Brian's [pseudonym] Homework: Brian's presentation is entitled "The Science of Love" and is a discussion on what causes love. He chose the topic because he was Googling "chemistry.com." It turned out to be a dating site and caught his attention. He mentioned that only seven percent of attraction between humans is due to what we say—the rest is believed to be chemical reactions in the body. He mentioned a lot of chemical terms such as serotonin, dopamine, oxytocin, and adrenaline. He explains that serotonin is found in the gastrointestinal tract and the central nervous system and is related to feelings of happiness. Dopamine is found in the brain and is related to reward-driven learning. Oxytocin, also called the "love hormone," has a role in sexual reproduction. And, adrenaline also plays a part in the science of love. Mrs. Waters adds more details to the conversation.

Brittany's Homework: Brittany's homework presentation is on "fermentation" (see fig. 3.3). She has studied the chemical reaction of yeast in rising dough. She has brought in a large tray of a delicious looking pizza (I can't wait to try it and am now working myself closer to her table). Brittany has made a poster upon which she has included several of the photographs she took during different stages of the baking process (see fig. 3.4). She also has prepared an extensive report on her homework assignment. Brittany is a culinary student who chose this topic because she had been working with dough in a cooking class the other day. I asked her how long she had spent on her homework assignment. She told me it had taken two hours after school yesterday to make the pizza and two hours last night to put together the poster. She took pictures with her iPhone and emailed them to herself so she could print them out. P.S. The pizza tasted great—I'm beginning to really enjoy this participant-observation research!

11:45 a.m. It's time for lunch and Mrs. Waters tells us that we will have to continue the homework presentations when we return from school vacation.

I have made a note in my journal to the effect that, so far, the flexible homework experiment has been quite successful. The presenters have all been highly engaged in the assignments they chose. They obviously spent quite a few hours on an assignment that had no time recommendations or constraints. And, several of them were "portfolio" quality. We have not had a single slacker yet! And, it seems that all of the students have had positive aesthetic experiences.

Moreover, I have been quite impressed with the level of engagement evidenced by all of the listeners in the class. This was especially apparent during the question and answer session following each presentation. Maybe the variety of topics has captured the interest of the audience. It stands in fairly sharp contrast to the traditional uniform "whole-class" assignment for which the teacher assigns only one topic to be done by all. All of the presentations for such an assignment tend to be the same.

March 2, 2011 (Day 14) Wednesday. Today, I have decided to visit Mrs. Waters' other two chemistry classes, Block 3 and Block 2 (in that order) to view more flexible homework presentations. My Chemistry Block 1 class has most of the higher performing students. But I was interested in assessing whether there were differences in the type or quality of the homework done by students who are at differing academic levels.

8:30 a.m. In Mrs. Waters Block 3 class today, there are only two presenters.

Michelle's Homework: Michelle's topic is "How Does Exercising Relate to Chemistry?" She has written a summary of an article she read and explains that she chose this topic because she does a lot of exercising. She tells us that she looked through three or four articles before she found one that was interesting. Unfortunately, I am not taking notes. However, as I look around the room, everyone else is engaged in her presentation. I am sure Michelle is doing a great job. For me personally, I find the subject of "exercising" to be repulsive and so I'm not really paying attention—I'm *aesthetically overdistanced.* But isn't that really the point of my research? What makes students engaged—is there a "hook" that Michelle could use to get me to focus on what she is saying?

John's (Block 3) Homework: John is calling his homework presentation "Burning Money." I'm already engaged because I can relate to this. It is "real world" stuff—I've been burning through lots of money recently. John has set up a demonstration involving a beaker filled with a solution of 50 percent alcohol and 50 percent water, some salt, a dollar bill, tongs, and a cigarette lighter. He first adds a pinch of salt to it, then tells us that the salt is used to change the color of the flame to make it more visible. John then soaks the dollar bill in the solution so that it is completely wet. Grabbing it with tongs, he shakes off any excess solution. Then, he lights the cigarette lighter and starts the dollar bill on fire. Amazingly, despite the presence of a large flame, it doesn't burn up. After the flame goes out, the dollar bill looks as good as ever. He explains that a dollar bill is made out of about 75 percent cotton so its material is actually closer to fabric than paper. It absorbs only the water. It is the alcohol that is actually burning, not the dollar bill. It stays wet and so remains intact. He chose to demonstrate this because it looked like fun. He has a friend who subscribes to a "YouTube" channel that features science experiments. John says he only spent ten minutes gathering materials. Mrs. Waters looked at me and says that she just ordered a lab kit for doing this same experiment.

11:15 a.m. Our Chemistry Block 1 class started with presentations.

Rosabel's Homework: Rosabel is presenting her homework entitled "Chemistry Related to Art." She chose this topic because she is a visual arts student and noticed her peers were choosing presentations related to their art. She discussed different types of acid etching. She spent about 30 minutes on the assignment.

1:15 p.m. I've gone to Mrs. Waters Chemistry Block 2 class to observe their flexible homework presentations. Again, this class is not quite at the same academic level as my Chemistry Block 1 class.

Cindy's Homework: Cindy has constructed a volcano. She has a 2-liter soda bottle placed on top of a baking pan. She has decorated the top of the bottle with a paper cone so that it resembles the shape of a volcano. She mixes three tablespoons of flour with three tablespoons of baking soda and pours the mixture into the bottle using a paper funnel. She then adds 60 drops of red food coloring and 18 drops of soap detergent into the bottle. Next, she pours 1½ cups of vinegar into the bottle. The mixture starts to foam and comes out the top. After the foaming stops, Cindy pours another 1½ cups of vinegar into the bottle. She explains to us that the chemical reaction that occurs when the vinegar comes into contact with the baking soda creates carbon dioxide gas that then expands and forces the mixture inside the bottle to foam out the top. I was surprised that the chemical reaction didn't appear to be particularly violent. Cindy says she chose this homework topic because she thought it would be cool and she spent about 1½ hours preparing for it.

Brianna's Homework: Brianna's topic is on the "Tin Can." She says she was looking up the chemistry term "mixture" online and came across a description of a tin can. She explains that nowadays, tin cans are not really made purely of tin—they are often tin-plated steel or tin-plated aluminum. Tin cans are used mostly for storing preserved foods and the tin plating prevents rust. Briana talked for about five minutes. She tells us that she spent about 30 minutes on her homework research and presentation.

Lyn's [pseudonym] Homework: Lyn has chosen a *Science World* magazine article about the element "lithium." However, in a short time, after Lyn is stumbling over some words and appearing not to understand what she is reading, Lyn admits to Mrs. Waters that, right before class, she had asked reading specialist Ms. Fox to find an article for her. Mrs. Waters tells Lyn she'll have to re-present.

A few words here:

Lyn's actions supports my hypothesis that, when given a simple research assignment, too often, low performing students will put in a minimum amount of effort, if any. They are likely to grab the first thing they come across—and often don't really understand what they are handing in. Further, they are not affected by peer pressure. In other words, the fact that someone else in the class may have submitted superior work does not seem to motivate them to intensify their efforts.

In contrast to this, higher performing students will spend time searching for something of interest (i.e., engaging). They often consider several possibilities before settling on one that they understand and are comfortable with presenting. And, they generally tend to be more competitive. They will measure their work against that of their peers.

March 4, 2011 (Day 16) Thursday, 9:50 a.m. (I was absent yesterday, March 3 (Day 15),—attended a NWEA workshop in Haverhill, Massachusetts). Jessica P. was sitting next to me today. I asked her if she presented yesterday. She tells me she did and her talk was on a YouTube demonstration of a coke bottle rocket. The rocket is created by adding five-to-seven "Mentos" mints to a 2-liter bottle of Diet Coke. A chemical reaction causes carbon dioxide gas to expand inside the bottle and then create an explosion. Jessica P. said she spent about five minutes selecting her topic because she had already seen this done in her last high school chemistry class.

March 7, 2011 (Day 17) Thursday. I'm making a note in my journal to reword the definition of the "Mandatory-Flexible" homework type to include a recommendation that a "flexible" assignment topic not be as broad as "something related to chemistry." Instead, it should be a more restrictive or focused topic, e.g., "you can design your own homework assignment, but it has to be related to 'exothermic and endothermic'." Mrs. Waters' first experiment with a flexible homework assignment was in fact to "design something related to chemistry." However, it was assigned in the first days of the semester, before the students had learned any significant amount of chemistry.

I am also making a note to recommend that Flexible Homework not be assigned too frequently. Students can get burned out or run out of innovative ideas.

April 15, 2011 (Day 46) Friday, 8:30 a.m. (The day before our week long April vacation). Mrs. Waters assigns a Mandatory-Flexible homework assignment to be completed over vacation. It is part of the preparation for the Rhode Island state science assessment (i.e., NECAP) in May. The assignment varies by student based on their most recent NWEA Science scores. Each of them will design their own homework based on their weakest NWEA RIT bands (i.e., assessment scores) from the three assessed areas (i.e., Earth and Space, Life Science, or Physical Science)—but it has to be linked to a specific GSE (Grade Span Expectation) of the state science standards. The NWEA science assessment is aligned with the Rhode Island state standards, so we use it as an indicator for potential student performance on the state's NECAP. Mrs. Waters had already requested that I (in my role as Dr. Butler, not chemistry student, John) provide each student with a report of their NWEA Science RIT scores so they can choose their assignment based on their weakest area (supplying NWEA results to students and teachers is one of my many tasks as the school's Data Analyst).

April 25, 2011 (Day 47) Monday, 8:30 a.m. (The day after our week long April vacation). Chemistry Block 1 begins with a POD (Problem of the Day) related to the NECAP science assessment. Looking at Mrs. Waters' agenda for today's class as it is written on the board, I see we have several activities to complete before we get to any of our flexible homework presentations. Our graded lab reports are now being returned together with a quiz we took before vacation.

9:25 a.m. Mrs. Waters tells us that we have time for a couple of presentations. Siraj volunteers to go first.

Siraj: She tells us that her weakest NWEA science area is *Life Science*. She found a YouTube video on photosynthesis entitled "Photosynthesis, let's get into this." It's a very cute video using a rap song to describe the photosynthesis process. Siraj chose the video because she found it interesting. Everyone in the class agreed that it was a well-done video and engaging.

Brittany: She tells us that her weakest NWEA science area is *Scientific Inquiry*. Brittany has created a PowerPoint presentation on how to make "rock candy."

Mrs. Waters announces that class is over and that we will continue with projects tomorrow.

April 26, 2011 (Day 48) Tuesday, 1:15 p.m. Our Chemistry Block 1 flexible homework presentations continue:

Katie: Her weakest NWEA science area is *Life Science*. She has written a paper on stem cells.

Sima: Her weakest NWEA science area is *Earth and Space Science*. She found a YouTube video about the moon entitled "Fun Science: Moon." She chose the video because it has a cute boy with a British accent doing the narration. All of the girls in the class agree with Sima that the boy is cute and want to know if he has made other science videos. With regards to the science content covered and quality of presentation, it seems quite good.

Manar: Her weakest NWEA science area is *Earth and Space Science*. She found a video entitled "A Star is Born" in which the formation of stars is discussed. She chose it because she likes the word "nebula."

Sarah: Her weakest NWEA science area is *Earth and Space Science*. She found a video from "Space School" about the Sun.

Lindsay: Her weakest NWEA science area is *Earth and Space Science*. She presented an article about global warning.

Mrs. Waters announces that class is over and that we will continue with presentations tomorrow.

April 27, 2011 (Day 49) Wednesday, 11:15 a.m. Our Chemistry Block 1 flexible homework presentations continue:

John: I presented first today. I didn't have any NWEA data on myself so I was free to select from any of three assessed areas. I chose Physical Science and my topic was entitled "A Proposed New Efficient Method for Determining the Electron Configuration for the Elements." I made a poster and wrote a paper on what I thought was an innovative method and one more efficient than what has traditionally been taught. During the process, I had an aesthetic experience—an "Aha!" moment—when I thought I may have discovered a unique connection to prime numbers. A detailed discussion of this presentation and experience is included in the appendix.

Brian: His weakest NWEA science area is *Life Science*. He tells us that he was looking online for something related to life Science and found a video on the "Brain and Love." Ruby notes that Brian chose a similar topic for a previous flexible assignment. Brian responds that it's just a coincidence.

Hector: His weakest NWEA science area is *Earth and Space Science*. He found a short video on YouTube entitled "The Night Sky Flash Sept 26 2008" by TheNightSkyGuy. The video discusses the glow of the Zodiacal Light from space debris.

Sarah: She has already presented but asks Mrs. Waters if she can present again. She has located another video from the cute British boy that Sima had found. The YouTube video is an excellent short video entitled "Fun Science:

Light." Again, all the girls in the class seem quite engaged. Hector and I are the only males in the class and I think I can speak for him. He and I are engaged in the content of the video and not having the same aesthetic experience as the girls!

12.45 p.m. End of presentations and off to lunch

1:15 p.m. After lunch, I decide to check out Sima's and Sarah's YouTube videos with the boy with the British accent to see what other science related videos he is in. A very quick internet search reveals that the boy's name is Charlie McDonnell. He is a 21-year-old English Video-Blogger and Musician from Bath who currently resides in London. As of the date of this online search, it appears that he has made well over a hundred videos and received over a million views. He is known on YouTube as "charlieissocoollike" (or just Charlie) and is currently the most subscribed YouTuber in the United Kingdom. I watched a few of the science related videos and found that the content he covered and presentation he made were done quite well and appropriate for a high school level science class.

May 9, 2011 (Day 57) Monday. Mrs. Waters assigns a Mandatory-Flexible Project assignment due on Friday. It, too, is geared toward NECAP preparation. All of her previously, flexible assignments have been in the form of homework. This is our first *Flexible Project* assignment. Mrs. Waters' Mandatory-Flexible Project: "Make a model of anything that represents something in Earth and Space Science, Life Science, or Physical Science linked to the science GSEs" (Grade Span Expectations—Rhode Island's state standards).

Just as there was a need to develop school-wide guiding principles for homework, there was a similar need to develop guiding principles for projects. Answering "What constitutes a project?" was the main focus of the research I conducted in developing the school's *Guiding Principles for Projects*. By pure coincidence, I developed eight guiding principles for them—the same number as I had arrived at for homework. The first six are listed below. The last two had to do with portfolio requirements.

1. *For purposes of establishing a school-wide definition,* a *project* is a deliverable and the result of either individual or collaborative effort. This is the culmination of a process that challenges students beyond a normal classroom or homework assignment.
2. *The deliverable must have tangibility*. In the case of an oral presentation, theatrical performance, or similar event, the deliverable allows for a review, e.g., an audiovisual recording. It is not a general critique.
3. *In addition to the question of tangibility, a project is not an assessment of student behavior*. In other words, it is not a grade for attendance, behavior, or participation.

4. *Projects meeting the tangibility requirement include, but are not restricted to, components such as*: an essay; a general research report; a lab report; a grouping of several assignments (e.g., multiple worksheets); a PowerPoint presentation; a construction of a model; a script; a film; a creation of a culinary piece (e.g., meal, dessert, appetizer); a creation of theater production (e.g., skit, play); a creation of a visual art piece (e.g., drawing, painting, sketch, sculpture).
5. *To the extent that projects are intended to be more demanding than normal classroom and homework assignments*, they carry a disproportionately greater weight in terms of school-wide grading policy.
6. *Whenever possible, projects should be personally relevant and customized to promote student ownership:* Students need to approach them with a positive attitude. They should be designed to give them choices and create opportunities for them to personalize their work. They should also give them more control over what they learn, how they learn it, and how they show that they have learned it (Vatterott, 2007). Students are more apt to be engaged if a project is aesthetically pleasing (e.g., relevant, interesting, organized and neatly presented).

May 12, 2011 (Day 60) Thursday, 6:00 p.m.–10:00 p.m. I am working on my flexible project assignment. I have decided to build a "simple" motor and have found a lot of interesting videos on YouTube. A few of these require only a battery and a paper clip or wire. However, I'm not having any success with "simple" motors and I'm getting a little frustrated. I scrap this idea and decide to build a small electric generator that will light up a small flashlight bulb. I will have to purchase some heavy-duty magnets, lots of magnet wire, and a few other items. I dismantle (i.e., break) my pencil sharpener so that I can use the crank arm.

May 13, 2011 (Day 61) Friday, 8:30 a.m. We are supposed to start presenting our flexible project assignments, but I am not ready to present today. However, when I get to class, I find that Mrs. Waters is out sick so the presentations will have to be put off until Monday. I'm thinking that this might be a good thing—I can have the entire weekend to work on my project. How many other students were thinking the same thing?

5:00 p.m.–12:00 midnight. I stopped by RadioShack on my way home from school to pick up some materials for this. I spent about thirty dollars out-of-pocket for some heavy-duty magnets, magnet wire, a light bulb, switch, and a voltmeter. I worked till midnight assembling my electric generator but still had a lot left to do when I went to bed.

May 14 and 15, 2011 (Saturday and Sunday). I spent 22 hours this weekend on my chemistry project in addition to the seven hours I had put into it on Friday. I went to bed at midnight on Friday; at 3:00 a.m. Sunday morning; and 2:00 a.m. Monday morning; so, I must have been quite 'engaged' in my work. Unfortunately, I ran into a lot of technical problems. After several tries, I did get the flashlight bulb to go on but it was very dim. So, I figured I needed more wire windings on the coil I had constructed. This meant several hours of rebuilding work. Finally, at around 1:00 a.m. Monday morning—SUCCESS! The bulb shone bright. Now, I could go to bed and try to catch four or five hours of sleep before I would have to get up to make the half-hour drive to school. But let me just light that bulb one more time before I go to bed. I turn the crank that spins the magnets within an electric wire coil. The faster I turn it, the brighter the bulb lights. I can turn it about 180 revolutions per minute, but I have designed a 9:1 gear mechanism so that the speed of the spinning magnets is actually about 1,620 revolutions per minute. This is quite fast and even a little violent. All of a sudden, the bulb goes dark. Apparently, my aggressive cranking and the resulting high speed have caused some screws to come loose and cut through the outer layers of the coil windings. This was disastrous as it had taken me about two hours to create the coil.

Simultaneously, in my frustration over this failure, when I lifted my contraption to assess the problem, I scraped and damaged a computer monitor with the sharp corner from a metal piece.

So, let's reflect on my experience thus far with Mrs. Waters' Flexible Project assignment. I have spent over thirty hours of my weekend time, over thirty dollars out-of-pocket, damaged my computer monitor, and destroyed a good pencil sharpener. I make a note in my journal: "Participant observation research can indeed have a dark side!"

I have no choice. Today in class, I will have to present my project as a "work-in-progress." It will be necessary to explain some of the successes and failures associated with it.

May 16, 2011 (Day 62) Monday, 7:55 a.m. Some students from Mrs. Waters' other chemistry classes are stopping by Room 205 before the 8:00 a.m. start of school to drop off their flexible projects.

Robert from one of the other chemistry classes, has created a large poster-sized painting on "The Life Cycle of a Star" (see fig. 3.5). He is one of the many gifted visual arts students at Beacon. Desiree, also from another chemistry class, has baked two cakes and decorated one as an animal cell and the other as a plant cell with important parts labeled (see fig. 3.6).

8:30 a.m. DEAR is over in Room 205. Our Chemistry Block 1 class begins. Mrs. Waters asks who would like to present first. I volunteer as I want to get this over with. I explain how my electric generator works, or rather, how it is supposed to work. Then I recount all of the technical difficulties I have encountered. I also explain that from my many years of experience in a few different careers, I have often learned more from mistakes (i.e., trial and error) than from immediate successes.

Following my presentation, the other students in Block 1 present. I briefly describe their projects here as a means of providing insight to the types of projects students chose in this flexible assignment:

Siraj: has created a beautiful painting of the human eye (see fig. 3.7). She has labeled the various parts of the eye and has typed up a paper relating to her model. She reads the essay to us.

Manar: She has made a three-dimensional model of the amino acid glutamine. She has also prepared a paper on it and reads it to us.

Ruby: She has created a large poster describing the making of a fruit battery (see fig. 3.8). In addition to several illustrations, she has included a list of materials and the steps of the procedure.

Katie: She has constructed a three-dimensional model of DNA out of "Play-Doh." She describes the different components of her model to us.

Sarah: She has made a cut-away model of the earth showing the four major inner layers: the *crust*; the *mantle*, the *outer core*; and, the *inner core* (see fig. 3.9). Her model is made from a large Styrofoam sphere. She has painted the outside of the sphere with oceans and continents to represent the surface of the Earth. In the cut-away section, she has painted and labeled the layers of the earth. Sarah has also written a paper on her project which she reads to us.

Sima: She has made a model of the human heart. However, she tells Mrs. Waters that she doesn't want to present today. She would like to write a paper to go along with her model and asks Mrs. Waters if she could present on Friday instead of today. Mrs. Waters agrees.

I find Sima's attitude very interesting because it's supporting my hypothesis: "higher performing students generally tend to be more academically competitive. They will measure their work against their peers." Allow me to defend my hypothesis: Sima has just witnessed several of her peers presenting their flexible project models. Even though she has constructed a three-dimensional model of the heart and is knowledgeable about what she will present, I propose that since most of the presenters so far have also written papers to go along with their models, Sima is thinking that she has not equaled their effort. For her own personal satisfaction, she has asked Mrs. Waters for the extension of time.

For the record, Sima has been sitting next to me almost every day in our chemistry class and I feel I know her better than any other student in the class. She often offers her thoughts out loud as she is doing her work. Also, based on conversations and observations to date, I have determined that she and I think a lot alike. I'm pretty confident that my perception of why she wants the extra time is correct. She has expressed a desire to pursue a career in medicine

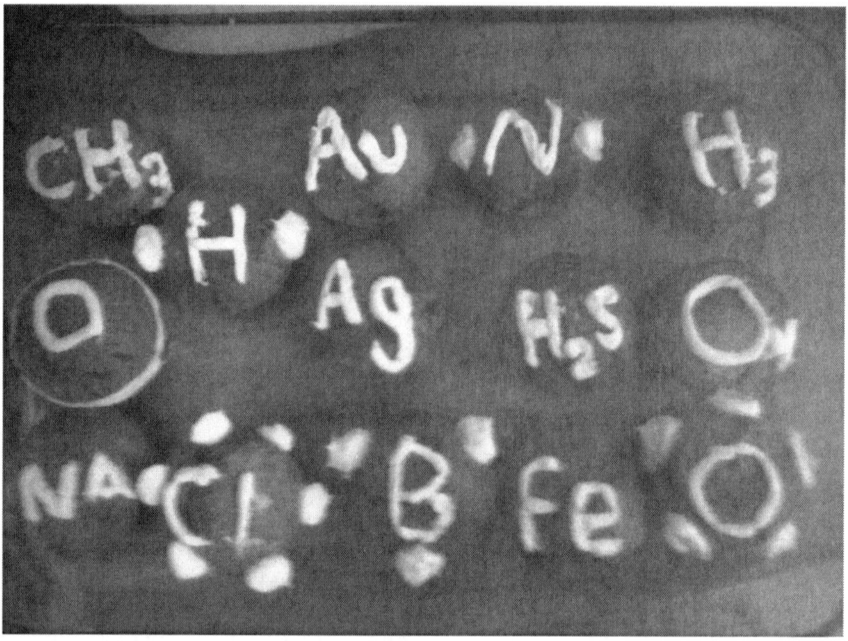

Figure 3.1. Sarah's Periodic Table cupcakes.

Figure 3.2. Katie's Periodic Table cupcakes.

Figure 3.3. Brittany's "Fermentation" pizza.

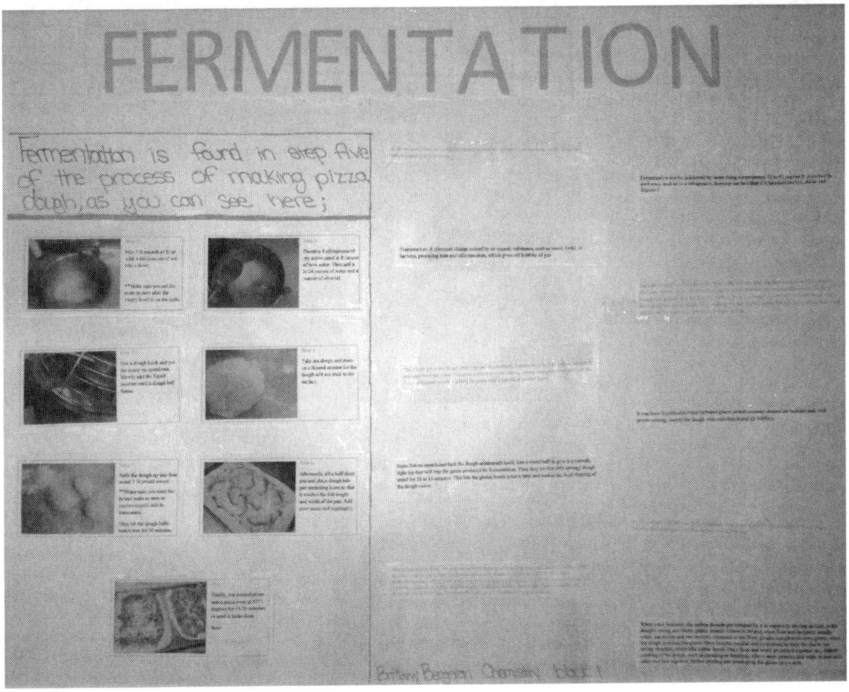

Figure 3.4. Brittany's "Fermentation" poster.

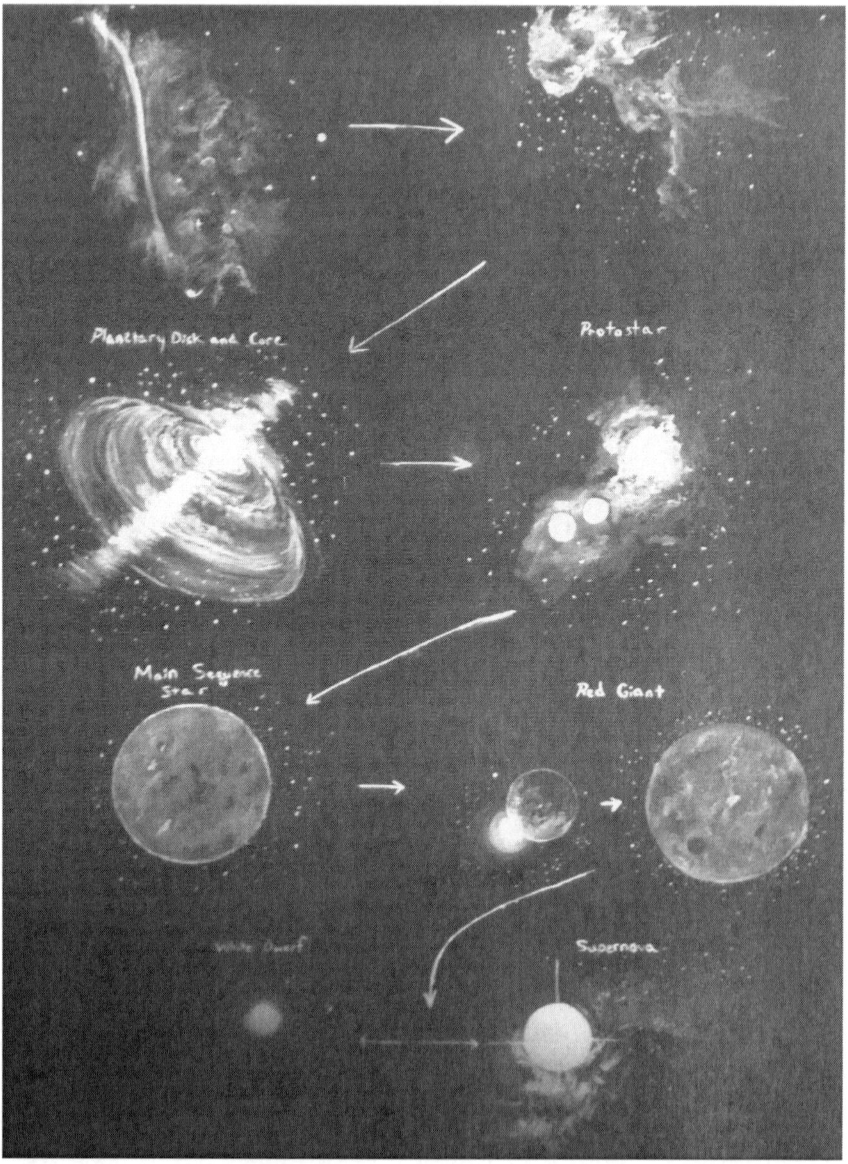

Figure 3.5. Robert's "The Life Cycle of a Star."

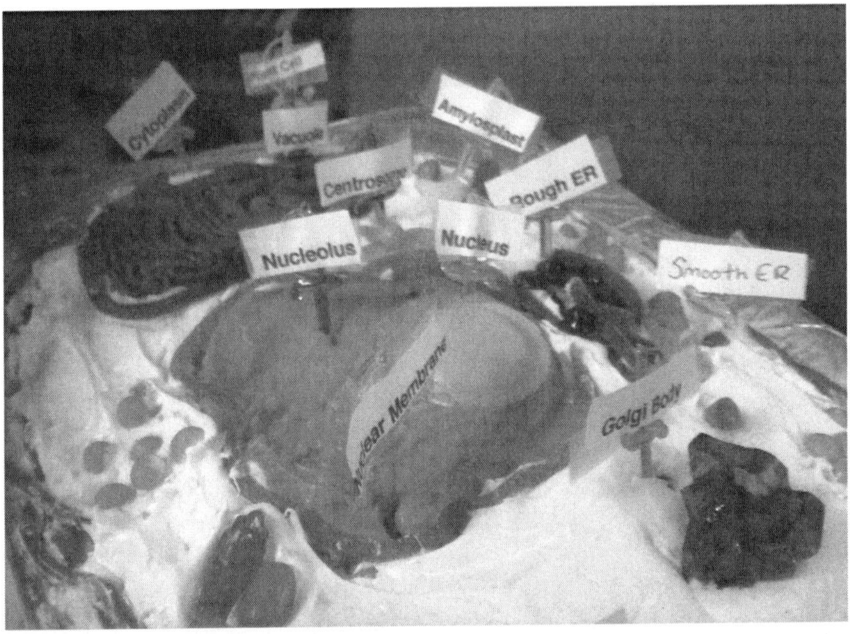

Figure 3.6. Desiree's "Plant Cell" cake.

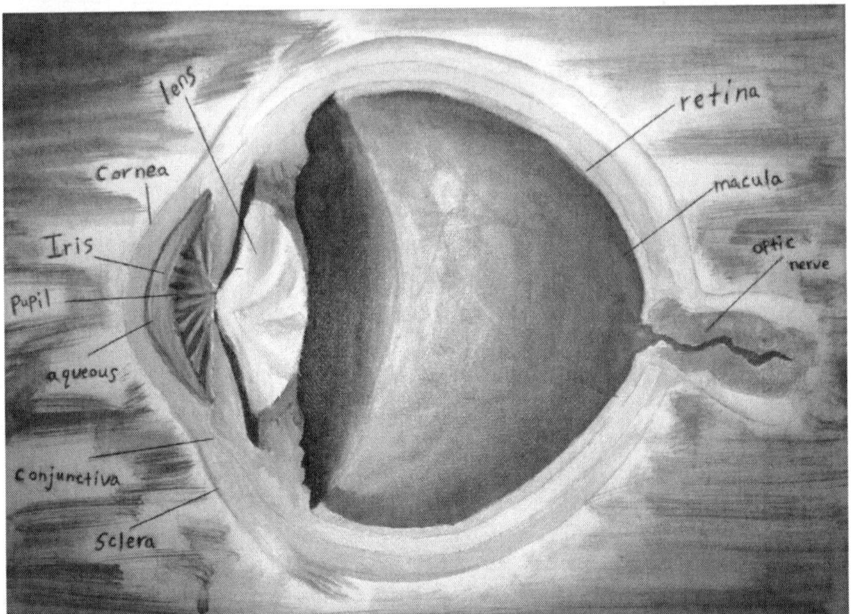

Figure 3.7. Siraj's "Human Eye."

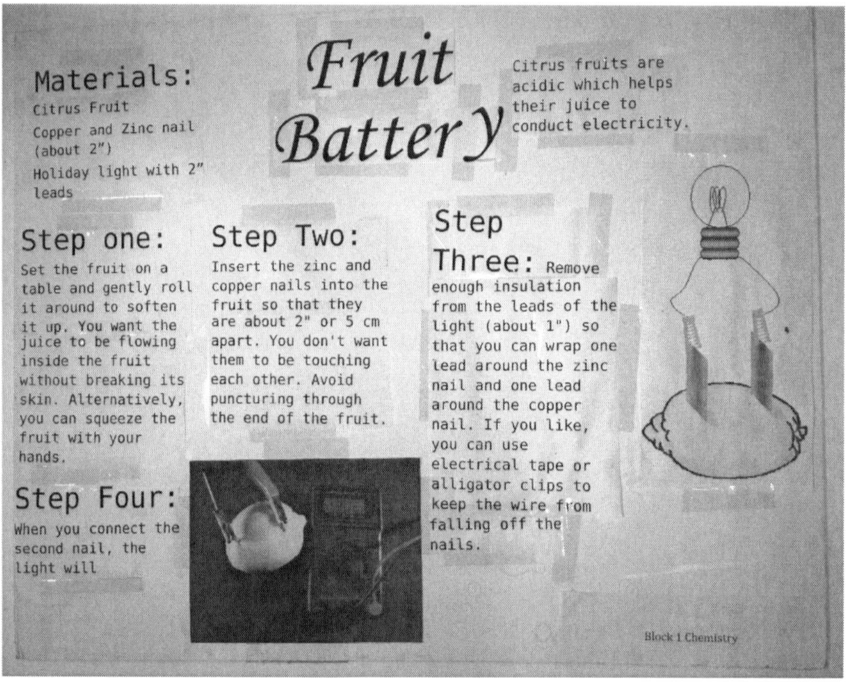

Figure 3.8. Ruby's "Fruit Battery."

and she takes her education very seriously—as does her twin sister Sarah. I continue to be impressed with the positive attitudes demonstrated by students in this higher performing chemistry class toward these flexible assignments.

Brian: He has made a three-dimensional model of the human digestive system out of "Play-Doh." It includes the abdomen, gall bladder, intestines, liver, spleen, and rectum.

Brittany: She has made a model of a cell and labeled all of its major features. She has also prepared a paper that she reads to us. Mrs. Waters' tells us that we will continue the presentations on Friday.

9:40 a.m. End of class.

May 17, 2011 (Day 63) Tuesday, 1:15 p.m. I sat with Mrs. Waters in the teachers' lunchroom. We talked about the flexible assignment that Robert created—the large poster of "The Life Cycle of a Star." He had presented it in his class yesterday and left it hanging in the classroom for the other chemistry classes. Besides being a great piece of artwork, students informed Mrs. Waters that one of the questions on the state science assessment dealt with the life cycle of stars. They believed they got it right because they remembered Robert's painting.

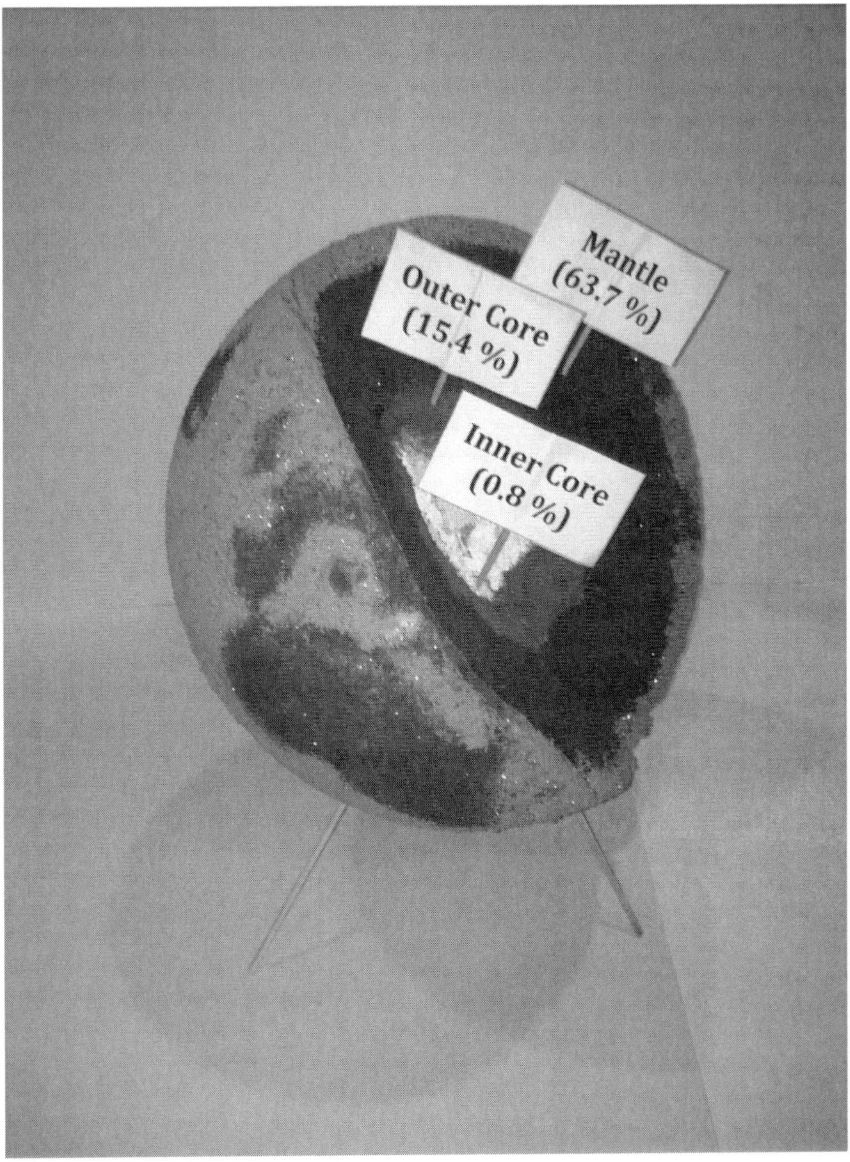

Figure 3.9. Sarah's inner Earth model.

As Mrs. Waters and I are talking, Mrs. Mowry (Beacon's Spanish teacher) is also participating in the conversation. We discussed how we could more clearly integrate the arts into academics. Actually, Mrs. Mowry, who was hired at the same time as I was (2005), has always been regarded as an expert at incorporating art into her lessons. We all agreed that we should more thoroughly explore opportunities for promoting student engagement in preparation for the state assessments and that we could do this by tapping into their art interests and native aesthetic sense.

Later that evening, I thought about this lunchroom conversation and about integrating the arts into other academic work. I reflected upon the many observations I had made in Mrs. Waters' class.

On Day 24, she had divided the class into six groups for the Atomic Structure Model poster activity. Each group had been assigned a different scientist to study, one of the several who had developed a model for the atomic structure: John Dalton, JJ Thompson, Ernest Rutherford, Niels Bohr, Louis de Broglie and Erwin Schrodinger. Each group had to research the work of one of these men and create a poster which contained an illustration of the model that was attributed to the scientist. It had to contain an explanation and list key features of the model. Then the groups had to arrange the posters in chronological order on the whiteboard in front of the class so that everyone could take notes off of the posters in preparation for a quiz that was to be given next Monday. Although this was an activity that related to the visual arts, I wondered if it could also have been presented from a theatrical perspective. How about if the six scientists had been invited to debate each other before an audience of peers—with each of the six defending their respective models?

I thought about the culinary connection: Sarah and Katie's cupcakes decorated as elements from the Periodic Table and arranged to form common chemical compounds; Brittany's pizza, the one she baked to study the chemical reaction of yeast; and, Desiree, who baked two cakes—decorating one as an animal cell and the other as a plant cell. And, the visual arts connection with Siraj's painting of the parts of the human eye, and Robert's painting of "The Life Cycle of a Star."

In many cases, there was a clear connection between the students' interest in art, their method of representing a scientific concept or principle, and their high level of engagement with their project.

May 20, 2011 (Day 66) Friday, 8:45 a.m. (The day after NECAP science testing)**.** Chemistry Block 1 student presentations continue from Monday:

Nicole: Has created several models of Bohr Diagrams. She chose this because she didn't understand the method when it was first introduced in class. She wanted to practice making the diagrams so as to improve.

Brian: Wanted to add more to his project on the digestion system, the one that was presented last Monday. He had done further research on the various components of it and presented additional facts related to the function and purpose of the liver, gall bladder, small intestine, and kidneys.

Rosabel: Created a three-dimensional working model of a Pulsar Star. The model lights up and when Rosabel rotates it, a blinking effect is created. It is representative of the blinking of such a star. She chose it because she wanted to do something related to Earth and Space science.

Lindsay: Created a poster on the human heart with illustrations of parts and purposes.

Sima: Has created a model of the human heart. She tells us that she chose the project because she had a heart-shaped potato at home. She has made a Styrofoam base with the heart shaped potato mounted on top. Recall that last Monday, Sima asked Mrs. Waters if she could have an extension so she could add a paper to her presentation. Sima read her paper to us today. Unfortunately, the potato has not done well with the one-week extension. It's now rotting and beginning to smell!

9:45 a.m. End of presentations. I make a note in my journal that the majority of flexible homework and flexible projects I have observed in Mrs. Waters' classes involved the deeper-thinking activities of Level Three or Level Four of Webb's DOK (Depth of Knowledge)—but not activities assigned by teachers, i.e., not the traditional way, but rather student-selected. Descriptors in these levels include terms such as Develop a Logical Argument, Use Concepts to Solve Non-Routine Problems, Investigate, Design, Connect, Apply Concepts, Analyze, and Create (see ch. 4, note 3).

May 25, 2011 (Day 69) Wednesday, A Conversation with Mrs. Waters: What have you learned about flexible homework? Mrs. Waters tells me that she found that the majority of her students spent much more time on their homework than she expected—even in her lower performing classes. She attributed this to the fact that they had the opportunity to link their assignment with one of their own areas of interest. Misty, for example, chose to do a flexible assignment on human organs because she hoped to work in the "special effects" industry someday.

A surprising number of students chose to use their artistic skill in creating their assignments. Culinary students baked cupcakes, cakes, and pizza; visual arts students created paintings and three-dimensional models. Mrs. Waters has concluded that integrating more of the arts into her science pedagogy will be a professional goal of hers in the future.

She also found that, when she asked students to align their choice of flexible assignment to a state science standard, they had no problem understanding the standard or creating an assignment that would make this possible.

Further, Mrs. Waters agreed with the "peer-pressure" argument of my hypothesis. She recalled that Sima, Brian, and Michelle (from another class) had all asked for extra time, even though all three had completed their assignments. In each case, their aim was to further improve their presentations.

Mrs. Waters also added that the extra credit opportunities that she often provides, which are always voluntary, continue to be a strong indicator of student engagement. She plans to continue to provide them frequently.

When I asked her about any disadvantages she had found with flexible assignments, she had an answer right away. She told me that, for her, the biggest challenge was to be equitable in grading them. They were all so different that it was difficult to establish a standard for evaluating them. Moreover, how does one accurately assess effort? It's much easier when all of the assignments are identical and comparisons between student work are easily made.

SOME CLOSING REMARKS ON THIS CHAPTER

It is quite obvious from my observations and from the evidence provided in the form of student work, that student engagement, particularly with regard to "time-on-task" and "depth of knowledge," both solely determined by the student, is significantly enhanced when students have a voice in creating their assignments. Linking the assignment with what is to them an area of interest also contributes to this. In terms of aesthetics, having a voice and linking their areas of interests, resulted in their approaching the assignments with passion! And, they were noticeably proud of their accomplishments. Further, ability to research, build models, and communicate in correct scientific terms are all twenty-first century science skills.

The finding that a significant number of Beacon Charter High School for the Arts students chose on their own to integrate their art into their flexible assignments is supportive of the STEM to STEAM[3] initiative in education.

> An approach to science that includes its aesthetic dimension could make artists feel more at home with science. They would then be able to see that themes such as form, order, and pattern are important in science as well as in art, that the creative experience is similar in the two areas, and that style and taste are relevant to areas outside of art. Then they could bring back to their art an enhanced perception of nature and of those who investigate it (Flannery 1992, 12).

In an effort to further explore student aesthetic experiences with these flexible assignments and other types of activities routinely assigned by Mrs. Waters, an exit survey was administered. It is discussed in chapter 5, "Mrs. Waters' Exit Survey: What does the research say?"

NOTES

1. *Living statue* refers to a mime artist or street entertainer who poses as a statue (i.e., immobile) for long periods of time. He or she normally dresses in realistic costume with makeup. Several Beacon student-artists perform as living-statues for many community events during each school year.

2. *At-risk,* where educational populations are concerned, refers to those students who are likely to drop out of school or be held back from moving to the next grade level. Determinants of this include: living in a single-parent family household, receiving welfare assistance, qualifying for free or reduced paid school lunch, living in a household where the primary language spoken is not English, and having a mother who did not graduate from high school.

3. STEM to STEAM. STEM is an acronym for Science, Technology, Engineering, and Math, and refers to initiatives designed to improve the education in these areas to meet the demands of the workforce. STEAM expands STEM to include an art component and refers to the initiative that promotes educational opportunities that foster innovation by linking the artistic creativity of artists with the thought habits of engineers and scientists.

Chapter Four

Mrs. Waters' Formal Lab Report

It doesn't make a difference of how beautiful your guess is. It doesn't make a difference of how smart you are, who made the guess, or what his name is. If it disagrees with experiment, it's wrong. That's all there is to it.

—Richard P. Feynman, Lecture at Cornell University, 1964[1]

COOKBOOK, GO PLAY OR SOMETHING IN BETWEEN?

The Olympics are totally engrossing, from the opening ceremony to the symbolic extinguishing of the Olympic flame on the final day. For the seventeen days of the London Games, NBC averaged 31.0 million viewers in prime time. Every day, and several times a day, we are given an update on the medal count for the participating countries. Even those viewers who do not tune in to a major broadcasting network will most likely get the update on their local news station or from other reporting sources. Of course, it's the gold medals that really count and American athletes do still generally perform well!

However, there are international competitions in which this is not the case, e.g., two major international assessments: (1) Trends in International Mathematics and Science Study (TIMSS)—which tests grades four and eight; and, (2) Program for International Student Assessment (PISA), conducted by the Organization for Economic Co-operation and Development (OECD) in the areas of reading, mathematics, and science, which tests 15-year-olds. The TIMSS studies are administered every four years. There was no significant difference between the US average science score at grade four in 1995 (542) and 2011 (544). But six countries outperformed the US fourth graders in 2011. Compared with 1995 (513), the US average science score at grade eight was 12 points higher in 2011 (525). However, in 2011, nine countries outperformed US eighth graders (Martin, Mullis, Foy, and Stanco 2012).

In his book, *Stem the Tide: Reforming Science, Technology, Engineering, and Math Education in America*, author David E. Drew states "American high school students have always performed at or near the bottom in international assessments" (Drew 2011, 25). In the 2009 PISA international assessment of 15-year olds, 30 countries outperformed the United States in math literacy, 22 countries outperformed the United States in science literacy, and 16 countries outperformed the United States in reading (OECD 2010).

The weak performance of US students compared with those of other countries has been an impetus behind the standards movement. The *National Science Education Standards* (NSES) were released in 1995 by the National Research Council (NRC). Developers of these standards focused primarily on inquiry. Since the release of this data nearly two decades ago, the research literature on "inquiry-based" teaching and learning has grown exponentially. With regard to science, the term "inquiry" may be defined as "the process of defining and investigating problems, formulating hypotheses, designing experiments, gathering data and drawing conclusions about problems" (Trowbridge, Bybee, and Powell 2000, 207).

As this book is being written, the *Next Generation Science Standards* (NGSS) are being developed. This is a joint effort between the National Research Council (NRC), the National Science Teachers Association (NSTA), the American Association for the Advancement of Science (AAAS), and Achieve. Table 4.1 provides a comparison of the list of essential practices for scientific inquiry which were found between the 1995 NSES and the 2012 NGSS draft. It is clear that "inquiry" will remain prominent in science education. One of the best opportunities for inquiry-based learning is in the science classroom, specifically in laboratory activity.

In their article "Modifying Cookbook Labs," authors Clark, Clough, and Berg (2000, 40) describe a continuum of student learning activities. At one extreme, they passively follow cookbook laboratory procedures (i.e., choreographed recipes); and, at the other, they create their own experiments—they investigate a question of their own choosing. "These extremes miss the large fertile middle ground that is typically more pedagogically sound than either end of the continuum (Lazarowitz and Tamir, 1994)." At the end of this course, where would I place Mrs. Waters' labs on the laboratory experience continuum? "Cookbook," "go play," or somewhere in between?

Table 4.1. Evolution of the Essential Features of the Standards

Essential Features[a] *National Science Education Standards (NSES) (NRC 2000)*	*Essential Practices Next Generation Science Standards (NGSS) (NRC 2012)*
1. Learners are engaged by scientifically oriented questions	1. Asking questions (for science) and defining problems (for engineering)
2. Learners give priority to evidence, which allows them to develop and evaluate explanations that address scientifically oriented questions	2. Developing and using models 3. Planning and carrying out investigations
3. Learners formulate explanations from evidence to address scientifically oriented questions	4. Analyzing and interpreting data 5. Using mathematics and conceptual thinking 6. Constructing explanations (for science) and designing solutions (for engineering)
4. Learners evaluate their explanations in light of alternative explanations, particularly those reflecting scientific understanding	7. Engaging in argument from evidence
5. Learners communicate and justify their proposed explanations	8. Obtaining, evaluating, and communicating information

[a] The term "features" in NRC (2000) has been changed to "practices" in NRC (2012)

A SPECIAL LAB DAY

Recall the earlier remark, "Mike Skeldon say's I won't last 10 days and Dr. Pilkington has given me 30 days." Thursday, March 24, 2011 (Day 30) is therefore significant for two reasons. First, as concerns my proposal to attend and participate in every one of Mrs. Waters' Chemistry Block 1 classes, I was about to disprove of some of the local naysayers.

Secondly, today was lab day and this was going to require a formal lab report. It would be worth a "project grade." Mrs. Waters' had been telling us from the start of the semester that we would be doing several labs, one of which would require a formal lab report. It would be equivalent in rigor and required format to a first-year college chemistry course.

Having signed in at 7:19 a.m. and proceeded to my office, I was hoping to get my chemistry "Pre-Lab Activity" homework assignment done. It was due in a couple of hours! This was an important assignment. The homework worksheet described the lab activity to be performed and discussed all of the associated safety issues. Mrs. Waters requires every student to complete the worksheet in order to qualify to participate in the lab. It is part of a learning strategy called *flipped teaching* or *flipped classroom*. Students preview new material the night before a class and then apply what they learned in a problem-solving activity in class the following day.

Some of my colleagues have stopped by to chat before the start of the first class. So, the pre-lab homework will have to wait. I also forgot that representatives from RIDE (Rhode Island Department of Education) would be here shortly for a de-briefing session on our progress with the new Rhode Model Educator Evaluation System. As noted, Beacon was one of five schools in the state selected for the pilot program. The meeting began at 8:15 and was still going on when I had to excuse myself at 9:25 to get ready for chemistry class.

I now have 15 minutes to complete the pre-lab homework assignment. I have already skimmed the reading of Mrs. Waters' handout "Just Breathe: The Oxygen Content of Air" which she adapted from a *Journal of Chemical Education* (JCE) classroom activity (JCE 2001), but have not yet begun the associated "Pre-Lab Activity" worksheet. Looking at the worksheet in more detail, I realize that there are two sides that need to be filled out; and, I will not have enough time to complete it. Mrs. Waters will not be collecting the homework right away. However, she will be coming around with her grade book and will probably flip the worksheet over to make sure everything is filled out. I'm hoping she won't have time to look in depth at the answers (she will do that later). Thus, the immediate goal for me is to try to fill the two pages with something that looks complete but does not fully answer the questions that are being asked. In other words, I need to focus on breadth not depth. If time permits, I'll go back and add the deep thinking pieces.

On the front page of the worksheet, we are asked to outline the procedure for the investigation by making a flow chart. In the process of drawing the picture, we must read about and understand the order of events that will occur in the actual lab experiment.

But, the procedure for this particular lab is quite long and I am by no means able to provide a detailed response. I am definitely feeling pressured. I flip the worksheet over and write two sentences in response to each of the two questions asked. I intentionally write in large script so that I can use up as much white space as possible. Then I grab my backpack and run to class.

Mrs. Waters has already rearranged the furniture into nine lab stations to facilitate the use of them by two students per station. On the board, she has written:

What will we be doing? 3/24/11

1) Check pre-lab homework
2) Do Percent O_2 Lab
3) "I Am" activity

Objectives: students will be able to

Determine the percent of O_2 in Room 205 using experimental methods [She often used abbreviations in her board notes: O_2 = oxygen]

Mrs. Waters gets everyone settled as quickly as possible so she can assign lab partners. This was standard procedure for every activity we did in chemistry—we never knew with whom we would be working. Knowing this was a "high stakes" assignment that required collecting reliable data during the experiment, I am selfishly hoping to be partnered with the best student possible. I silently begin prioritizing names in my mind, students who would make up my "dream" team. But one by one, the people I had chosen were assigned to other teams by Mrs. Waters. All the really smart, engaged students are being paired up.

Having exhausted the class roster less one student, Mrs. Waters finally says "Ashley [pseudonym] why don't you work with John." Well, I have to admit that evil thoughts towards Mrs. Waters went through my mind. Ashley is an extremely likable student but she's absent a lot; and, when she is in class, she has trouble staying focused. She rarely makes it through an entire class without getting out of her seat for some reason or another. And cognitively, she's an ESL student—that's *E*arth as a *S*econd *L*ocation. But, she is truly a sweet girl, always respectful to me, and Guidance tells me she has been dealing with many problems at home—a factor which is all too common among many of our students.

I begin the lab entitled "Just Breathe: The Oxygen Content of Air" with Ashley. While I prepare my notebook to record the data, Ashley weighs a beaker filled with water on the triple-beam scale at our station. She tells me it weighs "155 grams." I begin to record this but find myself thinking that I ought to check the measurement myself. Good thing! Ashley has defied the intentions of scales' engineering. There are well-defined equally spaced grooves machined along the length of each beam. The idea is to slide weights on the respective beams, locking them into the appropriate groove so that the scale is perfectly balanced. However, she has ignored the grooves and haphazardly positions the weights between them–on all three beams. I now proceed to explain to her how the scale works. The future of this undertaking immediately becomes clear as Nostradamus whispers in my ear: "Let Ashley record and you do the measuring." But I do make a note in my journal to suggest to Mrs. Waters that the science curriculum in earlier grades (e.g., ninth grade) should include a unit on various science lab equipment and discussion of types of instruments, how to use them, and maintenance.

The next steps in the procedure require taking readings of the height of the water in an inverted test tube every five minutes until it stabilizes. So, Ashley makes a table and uses her cell phone clock for keeping time as I take the measurements. Or, at least that was our team's game plan until we had taken the very first reading. Just five minutes into the process, Ashley tells me she needs to ask Mrs. Waters if she can leave to call home. She explains to me that she needs permission to take her meds—now, I'm starting to understand!

I take both the measurements and the five-minute-interval recordings for the next 15 minutes until she returns. When I give her the task of washing out a test tube—she disappears for a half-an-hour. So, I complete the required three trials myself. However, not everyone in the class has finished all of the trials and Mrs. Waters says "for those that didn't finish today, you can finish tomorrow."

A BRIEF DESCRIPTION OF THE LAB EXPERIMENT

This lab experiment is a common activity used in high school and college chemistry classes based on the original work by Birk, McGrath, and Gunter (1981, 804-805). They were in search of a "safe, rapid, simple, accurate, and reproducible" (804) method of determining the percentage (by volume) of oxygen in air. The experiment involves the chemical reaction that forms rust (iron oxide). Briefly described, steel wool is allowed to rust within an inverted test tube in water. The air trapped inside decreases in volume as oxygen is depleted in the reaction with the iron (thus forming rust). This causes the water level in the inverted test tube to rise. The decrease in the volume of air should directly correspond to the volume of oxygen initially present in the test tube and a calculated value by experiment should be equal or nearly equal to the accepted value of 21 percent. To accelerate the process, the steel wool is "activated" with acetic acid. Students record height measurements of the water in the test tube at five-minute intervals until the water level stabilizes.

In Mrs. Waters experiment, students were required to complete three different trials, each with a different amount of steel wool. The data collected was then used to calculate the percent of oxygen in air as well as a percent error based on the accepted value of 21 percent. Mrs. Waters assigned students in groups of two or three in all three of her chemistry classes. Data was collected from all three classes and posted on her website (i.e., Digication) for the analysis component of the formal lab report.

THE FORMAL LAB REPORT

In addition to the handout, "Just Breathe: The Oxygen Content of Air" on her website, Mrs. Waters has posted the seven documents she has created as resources to aid students in writing their formal lab report: (1) "Pre-Lab Activity" worksheet; (2) "Writing a Lab Report" document; (3) "Laboratory Report Scoring System"; (4) "Laboratory Report Check List"; (5) "Conclusion Example"; (6) "Abstract Example"; and, (7) an EXCEL table summarizing the lab experiment data from Mrs. Waters' three chemistry classes. A brief description of each follows.

1. *The Pre-lab Activity worksheet:* Every lab done in Mrs. Waters' chemistry class was preceded by a homework assignment, namely, to complete the Pre-Lab Activity two-sided worksheet (see figs. 4.1 and 4.2). No student was allowed to begin the lab without first doing this. Mrs. Waters has designed a standard template for the worksheet. She can use this to customize forms where needed, for the specific lab experiment to be performed. It contains five sections—three on the front, and two on the back. The first section has twenty laboratory safety symbols arranged in a rectangular array, each of which is to be marked by a check mark if a particular safety symbol applies to the investigation. The second section requires the student to list two steps in the procedure of the investigation that need special care or caution. The third section provides space for a required flow chart which outlines the procedure for the investigation. On the back (see fig. 4.2), the fourth requires students to state two of their own questions about the scientific concept(s) involved in the investigation. The fifth section asks if the student understands the procedure. If the student responds "no," there is space provided to write out his or her questions. Finally, at the bottom of the page there is a place for the student's signature. This acknowledges that the student has read the lab and will follow the laboratory rules, safety procedures, and techniques as these are detailed in the investigation's procedure.

2. *Writing a Lab Report document:* Mrs. Waters has created a three-page outline of the formal lab report. She has provided a brief description of each of the nine required report sections in order of presentation: "Title Page"; "Abstract/Purpose"; "Introduction/Background Information"; "Materials"; "Procedure"; "Data"; "Data Analysis"; "Conclusions"; and, "Work Cited/Reference."

3. *Laboratory Report Scoring System:* She has developed a comprehensive point-based scoring system for grading the formal lab report. It is a universal framework, but can be customized to a given lab experiment. There are twenty-one criteria that address the nine major sections of the report (see figs. 4.3, 4.4, and 4.5). The system is designed for efficiency in grading. Feedback to students is limited to informing them on areas of strengths and weaknesses. This is based on points earned out of the number of points possible. From this, students can refer to Mrs. Waters' extensive support documents which contain expectations and examples. She has handed them out and posted them on her website.

Each criterion of the lab report scoring system is subdivided and assigned a point value based on how well a requirement is met. For example, if there is no Title Page, no points are awarded. If the Title Page is complete and the format correct, the maximum (four points) is awarded for that task. Mrs. Waters has also weighted the criteria. The maximum possible points for the report varies depending on the particular lab experiment. (e.g., for all tasks of the Percent of Oxygen in Air lab report it is 137).

The document's title, "Scoring System," distinguishes the assessment tool from a *rubric*[2] that would be constructed with labels for proficiency-level labels (e.g., Above, Proficient, Below) and a brief description/scenario relative to the respective performance level. Mrs. Waters uses lots of rubrics but, for the formal lab report, she feels that her support documents for each of the nine sections of the report, together with her scoring system, are more reliable, effective, and efficient.

4. *Laboratory Report Check List:* Mrs. Waters created this document as a type of graphic organizer to help students, especially her special education students, organize the information required for each section of the lab report prior to writing it.

5. *Conclusion: An Example:* This document describes how to write a good conclusion. It provides examples of one that meets proficiency standards and one that does not.

6. *Abstract: An Example*: As with the conclusion document, the abstract document provides example of an abstract which meets proficiency standards and one that does not. It also describes how to write an abstract.

7. *EXCEL table of lab experiment data from Mrs. Waters three chemistry classes*: Unique to this lab is the inclusion of experimental data from all three of her chemistry classes. Students are required to use the three-class data set in their analysis and subsequent discussion and to base their conclusions on it. After all three classes had completed the experiment, I helped Mrs. Waters construct an EXCEL spreadsheet. She posted it online to be accessed by her students.

ESTABLISHING A BASELINE FOR MRS. WATERS' STUDENT LEARNING OBJECTIVES

One of the major changes and challenges for classroom teachers in recent years is the increased use of data in the instruction process. Mrs. Waters' Laboratory Report Scoring System is just one example of the several tools she uses to assess student learning in her classroom. Student performance data on this formal lab report will help her create a baseline from which she

Name ______________________________ Block _____ Date __________________

Laboratory Investigation Title: ____

• PRE-LAB ACTIVITY •

Laboratory Safety Symbols - Cautions

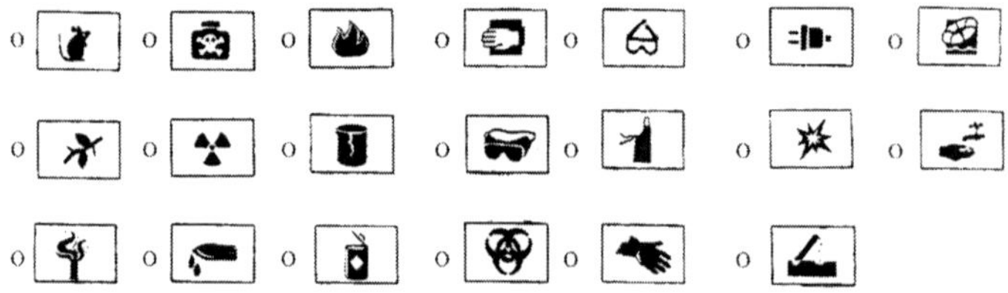

DIRECTIONS: Read the lab completely, then complete the following.

1. Identify the safety symbols pertaining to the investigation by darkening the circles before the symbol.

2. List two steps that needs special care or caution

   a) ____
   ____

   b) ____
   ____

3. Using a flow chart, outline the procedure for this investigation.

Flow Chart

Figure 4.1. Pre-Lab Activity worksheet—front side.

4. State two questions you have about the science concept(s) involved in this investigation.

Q1: ______________________________

____

____

____

Q2: ______________________________

____

____

____

5. Do you understand the procedure involved in this investigation?

O yes, I completely understand the procedure
O no, I am confused about a few steps in the procedure
O no, I am really confused with the procedure

If you answered "no" in question 5, write your questions below.

Q1: ______________________________

____

____

____

Q2: ______________________________

____

____

____

Q3: ______________________________

____

____

____

Q4: ______________________________

____

____

____

Safety Techniques & Procedure Acknowledgement

I ______________________________ have read the lab and will follow all laboratory rules, safety procedures and laboratory techniques as written in the procedure and presented during the Pre-Lab Activity.

Figure 4.2. Pre-Lab Activity worksheet—rear side.

Name ______________________________ Block _____ Date _______________

Laboratory Investigation Title: ______________________________

• Mrs. Waters' Laboratory Report Scoring System: Chemistry •

Task	Performance	Points	Points Earned
Title Page	o complete & followed format	4 points	
	o incomplete or did not follow format	2 points	
	o is missing	0 points	
Abstract Components (must include)	o purpose	2 points	
	o hypothesis	2 points	
	o hypothesis as an if, then, because statement	2 points	
	o **brief** description of techniques/procedure (0 points if too much written [i.e., not brief)	2 or 0 points	
	o major findings/conclusion	2 points	
	o voice of the 3rd person/past tense	2 points	
Abstract Quality	o is well written and logical; easy to follow	5 points	
	o needs to go into greater detail	3 points	
	o contains some inaccuracies/incomplete	2 points	
	o inaccurate; lacks logic/difficult to follow	0 points	
Background Information Topics (must include)	o (lab specific— customize here if applicable)	2 points	
	o (lab specific— customize here if applicable)	2 points	
	o (lab specific— customize here if applicable)	2 points	
	o (lab specific— customize here if applicable)	2 points	
	o (lab specific— customize here if applicable)	2 points	
	o (lab specific— customize here if applicable)	2 points	
	o (lab specific— customize here if applicable)	2 points	
	o (lab specific— customize here if applicable)	2 points	
Background Information Quality	o is well written and logical; easy to follow	6 points	
	o needs to go into greater detail	3 points	
	o sections are not in student's own words	2 points	
	o is confusing; lacks logic/difficult to follow	0 points	
Materials	o has a complete materials list	2 points	
	o has an incomplete materials list	1 point	
	o materials list is missing	0 points	
Procedure Completeness	o has a complete procedure	2 points	
	o has an incomplete procedure	1 point	
	o procedure is missing	0 points	
Procedure Quality	o is well written and logical; easy to follow	5 points	
	o needs to go into greater detail	3 points	
	o is confusing; lacks logic/difficult to follow	0 points	

Points earned this page (Subtotal 1 of 3) = [ ]

Figure 4.3. Lab Report Scoring System—page 1/3.

Task	Performance	Points	Points Earned
Data and Data Tables	o all data table(s) are complete and accurate	12 points	
	o data tables are incomplete and/or contain inaccuracies	8 points	
	o data is given but not in table format	6 points	
	o data and data tables are missing	0 points	
Data Table Components/ Quality	o is titled	2 points	
	o is labeled/ and with correct units	2 points	
	o is organized and neat	2 points	
Additional Data Related	o observations	2 points	
	o diagrams/drawings	2 points	
	o other: (customize here)	2 points	
Data Discussion Components	o a complete analysis of your findings	6 points	
	o an incomplete analysis of your findings	3 points	
	o the discussion section is missing	0 points	
References to Data	o clear references to the data collected	5 points	
	o reference data but lack clarity	3 points	
	o no reference to the data collected	0 points	
Data Discussion Quality	o is well written and logical; easy to follow	5 points	
	o needs to go into greater detail	3 points	
	o does not stay on focus topic	2 points	
	o is confusing; lacks logic/difficult to follow	1 points	
Graph Inclusion	o all possible graphs are included	5 points	
	o not all graphs are included	3 points	
	o no graphs are involved	0 points	
Graph Quality	o graphs accurately represent data	5 points	
	o graphs are constructed properly/format	2 points	
	o graphs are correctly explained	3 points	
Additional Analysis	o calculations are included	2 points	
	o percentage error is reported	2 points	
	o identifies problems in the investigation	2 points	
	o provides possible solutions to the problems	2 points	
	o suggests ways to extend the investigation	2 points	
Conclusion	o is scientifically correct/logical (2 points); if some inaccuracies (1 point); otherwise 0	2, 1, or 0 points	
	o identifies the major findings	2 points	
	o uses data as to support statements/findings	2 points	
	o is well written and easy to follow	2 points	
	o restates the hypothesis/topic sentence	2 points	
	o uses appropriate vocabulary	2 points	
	o uses voice of the 3rd person/past tense	2 points	

Points earned this page (Subtotal 2 of 3) = ☐

Figure 4.4. Lab Report Scoring System—page 2/3.

Task	Performance	Points	Points Earned
Literature Cited	o contains a literature cited page	5 points	
	o identifies three or more references	2 points	
	o is appropriately formatted	2 points	
Format	o report is typed	2 points	
	o follows formatting completely:	2 points	
	□ font size, script and color are appropriate		
	□ pages/sections were in the correct order		
	□ "art work" was not present on the title page		
	□ proper headings were used		
	□ columns/centering and line justifications were correct		
	o work is neat:	2 points	
	□ no white-out, smudges, eraser marks or crumpled/damaged pages		
Writing Mechanics	o does not contain any spelling errors	3 points	
	o contains no major grammatical errors	3 points	
	o is written in the student's own words	3 points	

Points earned this page (Subtotal 3 of 3) = ☐

Total Points Earned = Subtotal 1 + Subtotal 2 + Subtotal 3 = ☐

Lab Report Total Score = $\frac{\textbf{Total Points Earned}}{\textbf{Total Points Possible}} \times 100\%$ **= --------- x 100% =** ☐

Recommendations, Suggestions, and Comments for Improving Student's Work
Student needs to work on: o following directions/format o writing an abstract (review directions given) o writing a hypothesis o designing data tables o their research skills o providing greater details in their writing o using their data as evidence to support their statements o writing a conclusion (review directions given) o checking their spelling o checking their grammar

Figure 4.5. Lab Report Scoring System—page 3/3.

can identify their strengths and weaknesses and assess the effectiveness of her teaching.

As discussed earlier, a very detailed scoring system was used to assess each of the tasks that were required to create the lab report. Twenty-one distinct criteria were assessed, each with its own weighted point system. However, in the analyses, each score was adjusted to a percentage that had been gotten correct. Further, each criterion was assigned a "Depth-of-Knowledge (DOK) Level"[3] (Level One = Recall, Level Two = Skill/Concept, Level Three = Strategic Thinking, Level Four = Extended Thinking). Table 4.2 provides the resultant *means* and *standard deviations* (SD), minimum and maximum scores, along with the DOK levels for each criterion. All thirty-three lab reports received from Mrs. Waters' three chemistry classes were scored and factored into the analyses. The twenty-one tasks involved are arranged from lowest to highest mean score in the table. When they are placed in this order, an interesting pattern evidences itself.

The students' final chemistry class grade had a strong[4], significant, inverse correlation with the DOK level—the higher the DOK, the lower the final course grade, $r(31) = -.779$, $p < .0001$. Students were able to perform quite

Table 4.2. Results of the Laboratory Report Scoring System (n = 33)

Lab Report Task	*Mean*	*SD*	*Min*	*Max*	*DOK*[a]
Additional Data Related	34.091	35.804	.000	100.000	3
References to Data	42.424	31.921	.000	100.000	3
Graph Quality	42.424	39.610	.000	100.000	2
Data Discussion Quality	47.778	28.071	.000	100.000	3
Additional Analysis	49.091	33.575	.000	100.000	4
Data Discussion Components	50.909	29.194	.000	100.000	2
Literature Cited	56.970	43.909	.000	100.000	1
Graph Inclusion	60.000	43.301	.000	100.000	2
Data Table Components/Quality	60.606	32.760	.000	100.000	2
Background Information Topics	61.364	41.501	.000	100.000	2
Background Information Quality	63.131	41.616	.000	100.000	2
Conclusion	66.450	28.598	.000	100.000	3
Data and Data Tables	67.677	38.401	.000	100.000	2
Abstract Quality	77.576	21.655	40.000	100.000	1
Abstract Components	85.859	14.506	50.000	100.000	1
Procedure Quality	93.939	19.675	.000	100.000	1
Procedure Completeness	98.485	8.704	50.000	100.000	1
Title Page	100.000	.000	100.000	100.000	1
Materials	100.000	.000	100.000	100.000	1
Format	100.000	.000	100.000	100.000	1
Writing Mechanics	100.000	.000	100.000	100.000	1

[a] Depth of Knowledge (Webb 2005)

well on lower level tasks that involved the recall of information such as making lists (e.g., the materials required for the lab); describing a procedure; writing without spelling and grammar mistakes, writing in their own words (i.e., writing mechanics); following formatting instructions such as font size, proper headings, and typing text (i.e., format); and, recalling elements and details of story structure (e.g., using Mrs. Waters' examples for the title page and abstract).

However, the tasks of higher DOK Levels (3 & 4) were troublesome for many students. In particular, they struggled with inquiry-related tasks involving the analysis and discussion of data, with developing logical arguments, and with explaining phenomena in terms of concepts.

Other observations from the analyses of the thirty-three student lab reports include:

- Only 38 percent (8/21) of the criteria had an average student score of 70 or above—all eight of these being of DOK Level One.
- Of the 33 students submitting lab reports, 45 percent (15/33) received a grade below 70 (the cut-off for passing).
- The mean total score for the 33 lab reports was 69.5 (SD = ± 15.2). This was below what was expected given the amount of class time that had been committed to this and the extensive online resources provided by Mrs. Waters.
- The mean final course grade for the 33 students was 83.0 (SD = ±10.5).
- The students' final chemistry class grade had a moderate-to-strong[4], significant, positive correlation with the their lab report total score, $r(31) = .696$, $p < .0001$. A scatter plot of the total score for the lab report as a predictor of final course grade is shown in figure 4.6. The regression equation was determined to be:
  Final Chemistry Class Grade = 0.4819 x (Lab Report Total Score) + 49.493
  Note: As a simple check, substituting the mean score of 69.5 for the total score on the lab report in the regression equation results in a predicted final chemistry grade of 83.0. This matches the actual final chemistry course mean grade of 83.0 for the 33 students.
- No significant differences were found between gender (females = 27; males = six) as to performance on either the total score for the lab report or in their final course grade.

Creating the formal lab report is a rigorous assignment. It requires that students utilize skills learned across several content areas: reading, language usage, math, and software use, i.e., Microsoft Office. Given this fact, my analyses also considered the relationship between students' total score for their lab report, their final course grade, and their NWEA scores in science concepts, general science, reading, language usage, and math. These are assessments that are administered to every Beacon Charter High School; they are aligned to NECAP, the Rhode Island state assessments. Bivariate analyses were performed between the five NWEA assessed areas and the two chemistry class variables: "the total score for the lab report" and the "final course grade." The Pearson *r* correlations, along with their respective *p*-values, for the 21 analyses of interest, are shown in table 4.3. Significant findings were revealed for 67 percent (14/21) of the analyses.

Major findings from table 4.3 include the fact that significant relationships were identified between language usage, math, and reading, and students' final course grade. These correlations were found to be generally moderately strong and positive in nature. A similar pattern evidenced itself where the

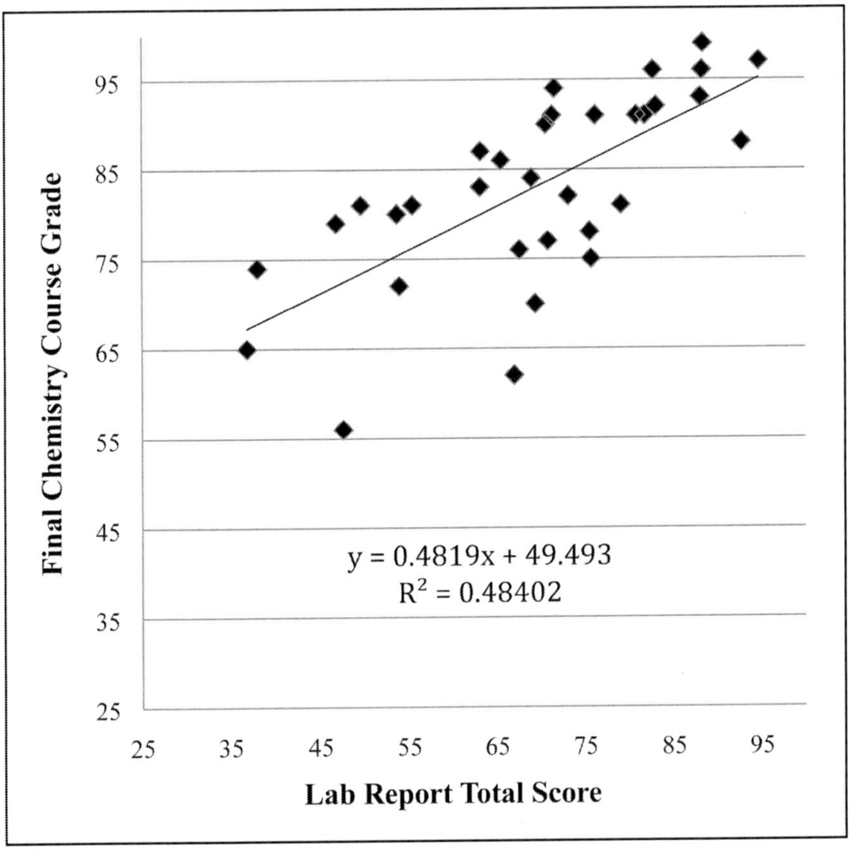

Figure 4.6. Predicting final course grade.

Table 4.3. Correlations r between NWEA Assessments, Chemistry Lab Report Total Rubric Score and Final Chemistry Course Grade

	NWEA Science Concepts	*NWEA General Science*	*NWEA Language Usage*	*NWEA Math*	*NWEA Reading*	*Lab Report Total Score*	*Final Course Grade*
NWEA Science Concepts	1.00						
NWEA General Science	.46[a] [.0072]	1.00					
NWEA Language Usage	.53 [.0016]	.43 [.0132]	1.00				
NWEA Math	.71 [< .0001]	.51 [.0025]	.70 [< .0001]	1.00			
NWEA Reading		.47 [.0060]	.81 [< .0001]	.78 [< .0001]	1.00		
Lab Report Total Score	.13 (.4739)	.10 (.5905)	.26 (.1461)	.32[b] (.0662)	.27 (.1358)	1.00	
Final Course Grade	.23 (.1958)	.16 (.3801)	.45 [.0090]	.37 [.0358]	.42 [.0139]	.70 [< .0001]	1.00

[a]r-values are shown followed by p-values in parenthesis ( ) or square brackets [ ]. Significance at the .05 level is indicated by the p-value in square brackets [p]. Approximate guidelines for describing the strength of r-values are found in Hatcher and Stepanski (1994): ±1.00 = perfect correlation; ±.80 = strong correlation; ±.50 = moderate correlation; ±.20 = weak correlation; and, ±.00 = no correlation.

[b]Considered "trend-level" significant: ($.05 \leq p < .10$)

relationships between language usage, math, and reading were concerned, with both "scientific concepts" and "general science." Again, correlations are generally moderately strong and positive in nature.

Somewhat surprising was the fact that there were no significant findings (at the $p < .05$ level of significance) relative to the five NWEA assessed content areas and the total score for the lab report. However, there was a "trend-level" significance ($.05 \leq p < .10$) found between students' NWEA math score and lab report total score ($p = .0662$).

Given that the total score on the lab report was found to be a predictor of final course grade, it is surprising that the former was not significantly correlated with language usage, math, reading, scientific concepts and general science. Further, the 33 students who submitted lab reports, were part of the grade-11 cohort ($N = 46$) that scored above the state averages on the NECAP mathematics, reading, and writing assessments. In addition, a look at the NWEA 2008 Normative Data revealed that the mean scores of the 33 students were "above grade level" in both reading and language usage and "at grade level" for math. Based on these assessments, it was reasonable to expect better performance on the formal lab report than what had been delivered. One interpretation of this may be that the process of creating the formal lab report (to this level of rigor) was new to students and involved skills that they had not yet fully developed and/or routinely practiced in other content areas. This provides an opportunity for Mrs. Waters to pursue interdisciplinary collaboration in the future.

SOME OBSERVATIONS AND COMMENTS FROM OTHER LAB DAYS

Thursday, March 17, 2011 (Day 25), 9:50 a.m. (*The Penny Lab*). The very first lab in Mrs. Waters' Chemistry Block 1 class is an experiment called "The Penny Lab." Briefly, it involves exploring *alchemy* by plating pennies with zinc and brass. On the board, she has written:

> What will we be doing? 3/17/11
> 1) Demo—green flame—excited electrons
> 2) Leprechaun Gold
> 3) Go over results
> Objectives: students will be able to
> – create an alloy
> – explain the chemical change that occurs between two metals and NaOH

For the experiment today, Mrs. Waters has arranged the classroom tables into five workstations. These are placed up against the windows in the southeast corner of the room. Each one is outfitted with a triple-beam scale, hot plate, tongs, spoon, beaker, and a watch glass. The windows above the tables are open to ensure that the fumes escape.

Before we begin, Mrs. Waters goes over the rules for safety in the lab. She emphasizes the requirement of wearing safety glasses at all times. She now checks that each of us has completed the "Pre-Lab Activity" homework assignment. Mine is done! As mentioned, part of the worksheet requires making a flow chart of the lab procedure. I did a lot of this during my engineering days—using common geometric shapes such as circles, rectangles, parallelograms, trapezoids, and triangles, strategically connected with lines and arrows. I notice that Siraj and Rosabel, both seated in front of me, have replaced the common geometric shapes with nice three-dimensional sketches of beakers and scales—a pictorial approach. I make a note in my journal that both of these students are "visual arts" students. Mrs. Waters places us in groups of four. My lab partners today are Hector, Lindsay, and Jessica G.

Before moving to the stations, Mrs. Waters conducts a demo. She creates a green flame using boric acid and methanol.

We then proceed to our lab station with our partners. Jessica G. tells us that she broke her glasses at home and can't see very well. She volunteers to be our scribe. Our group is working well together, carefully following the procedure. I notice Mrs. Waters circling the room, visiting each lab group to monitor its progress. I can hear her posing questions to the other groups. When it is our turn, she smiles and tells us that our results are exactly what she wants to see. She tells the rest of the class to check out the three very distinct colors of the pennies that our group has produced (see fig. 4.7). One is simply polished with steel wool to reveal its original shiny copper color. It serves as the control item; zinc plating on a second penny has apparently transformed its surface into shiny silver; and, by the further mysterious powers learned from alchemists, the brass plating of the third penny has created bright gold—at least it looks like gold, but I don't think we'll need to be escorted through the building by security because of this!

During the lab, the instructions call for setting aside the control penny. I place it under my notebook back at the table where I normally sit. When I go to retrieve it, I find Ruby [pseudonym] now sitting in my seat with her lab group at my table. My notebook has been placed on another table and there is no sign of my group's control penny. Ruby is her normal rude self and tells me to get another one (more evidence that I am viewed as a fellow student and not as a teacher). I eventually find the original one on Mrs. Waters' desk—she had picked it up from the floor. I have learned a valuable lesson in chemistry lab today. Labs and airports share something in common: "Please do not leave items unattended. Any unattended items will be removed."

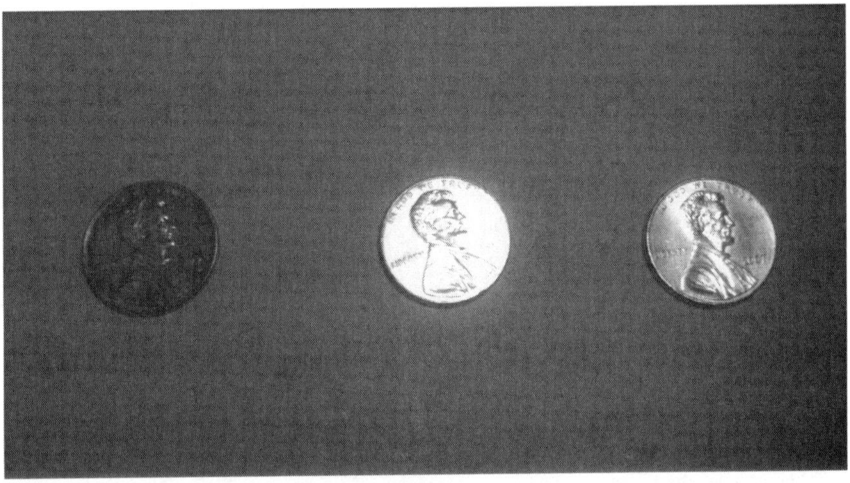

Figure 4.7. Copper, silver, and gold pennies.

All the lab groups seemed to be highly engaged and working well together. However, one of them has had a lot of trouble getting the expected results and Mrs. Waters asks us to work with them. She continuously walks around making sure that everyone is working safely. She herself has been pouring out the sodium hydroxide required in this experiment because it can burn and/or dissolve skin. Since we are nearing the end of class, Mrs. Waters instructs everyone on the cleanup procedure. Ten minutes before the bell she facilitates a whole-class discussion on our lab experience thus far. She assigns the homework to accompany it, namely, to research and answer the questions included on the lab handout.

With regard to student feedback on today's lab, before I leave class, I asked Sima what she thought of the lab. She had previously stated to Mrs. Waters that she wanted to do more "hands-on" activities. Sima said she enjoyed it. Later that day, as I was getting on the elevator, I saw Mary. I stopped and asked her the same question "What did you think of today's lab?" She quickly reached into her pocket and pulled out her shiny gold penny, proudly showing me the piece of gold and saying that she really liked the lab. Shortly thereafter, I ran into Nichole and also asked her how she had liked it. She said she had really enjoyed it, and, just like Mary, she pulled out her shiny gold penny to show me. Admittedly, I too, was proud of my group's results and had saved all three pennies. I think if I polled all of the students who had participated in it, there would be consensus on the fact that they had been engaged. I make a note in my journal that I should further investigate their feelings towards lab work in an exit survey at the end of the semester (see ch. 5).

2:30 p.m. The school day ends for students. I go to see Mrs. Waters. We often meet after school to discuss things from a teacher-to-teacher perspective. This is always a great opportunity to further explore her educational philosophies, to clarify observations, make suggestions, and collegially share ideas. At this particular meeting, I mention to her that earlier this morning, Mike Skeldon (Academic Dean and Principal in Residence) had asked me how many more days I intended to stay in chemistry class (I was now at day 25). I told him that I was going to stay to the end and that, amongst other things, I was learning a lot about student engagement—things that I would eventually be able to share with the rest of the faculty. Mrs. Waters then shared with me that Mike had asked her if I was influencing her teaching. She had told him "not at all"—only in the first few days was she aware of me. She further told him that even if she changed her teaching to put on a "dog and pony" show, no teacher could keep the act going this long. "After awhile, your true colors would show. You can only be a magician for so long before someone sees through the magic." We ended our discussion and I went back to my office, packed up my laptop, and headed home.

That evening, while I was working on the penny lab homework assignment, I discovered online that the composition of US pennies had changed several times over the years. For example, in 1943 (only), pennies were processed as zinc-coated steel because of a shortage of copper during World War II. I also found that the most recent change had been made in 1982, to 97.5 percent zinc and only 2.5 percent copper. So, I began to formulate a research question: "Do pennies minted prior to 1982 cause variation in outcomes in this experiment, as opposed to those minted after 1982? In other words, do the significant changes in their composition make a difference?" Then I suddenly had a thought: How about a "Test a Hypothesis Day" at the end of the semester. On such a day, students could perform their own lab experiment based on some hypothesis they had developed relative to any of Mrs. Waters' labs from the semester. I approached her with this suggestion and she thought it was a great idea. She mentioned it to our class the next day and everyone seemed enthused. Some students immediately had ideas about this and shared them with the class. Unfortunately, this project never came to fruition because of time limitations at the end of the semester. However, Mrs. Waters likely has plans for it in the future.

*Tuesday, April 5, 2011 (Day 38), 1:15 p.*m. *Electron Probability Lab.* This was our third of five labs. Normally, we would have performed it on a Thursday but because of some scheduling issues and the fact that it was a very simple one and did not require reconfiguring tables or dealing with chemicals, Mrs. Waters scheduled it for today. On the board, she had written:

What will we be doing? 4/5/11
1) COD

2) POD—PPT slide
3) Go over "electron in atoms" notes
4) Check H.W. Study Guide + Collider
5) Start Electron Probability Lab
Objectives: students will be able to
– Identify the inadequacies of the Rutherford Model
– Identify the new proposal in the Bohr Model of the atom
– Describe the energies and positions of the e^- according to the Quantum Mechanical Model

[NOTE: COD = Chemical of the Day; POD = Problem of the Day; PPT = PowerPoint; HW = Homework, e^- = electron.]

Of the total of 79 days I spent in chemistry class, Day 38 was perhaps the strangest. First, Ruby was 15 minutes late. The time required for her to walk from her visual arts to her chemistry class had gradually become extended—a pattern of continuous lateness had developed. She was always the last one to arrive in class. But today being 15 minutes late was unacceptable. When Mrs. Waters questioned her, Ruby said she was coming from art class. At present I am a student in this class. Nevertheless, my "teacher's" mind is about to explode because every Beacon student gets art class every day for their entire four years there. By contrast to this, they have half as many hours in science and math classes. Moreover, the State of Rhode Island's NECAP assesses reading, math, science, and writing, not art. Considering that Mrs. Waters is preparing the eleventh-grade students for the state science assessments and that she will be perceived as being directly responsible for their scores, I find Ruby's behavior, attitude, and subsequent interruption of the class, disrespectful. Mrs. Waters calls her aside and tells her she will be getting a zero for both the COD (Chemical of the Day) and POD (Problem of the Day) activities she has missed. We now go over the notes on electrons in atoms and, all of a sudden, three of the girls, Ashley being one, get up and leave the class without asking permission. Again, this is one strange class! Nothing like this has happened before. One young lady comes back, but the others have flown the coop. Mrs. Waters immediately calls the office to inform the Dean of Students of the incident.

After settling down again, we begin the Electron Probability Lab. Mrs. Waters assigns partners—mine is Mary. Mrs. Waters has Siraj pass out the procedure for this. It is a piece of paper with a target printed on it. The target is comprised of a number of concentric circles. She also hands out colored marking pens with felt tips. The lab is simple. We place the target on the floor and, from table height, drop the marker onto the target, aiming for the innermost circle ("the bulls-eye"). We are asked to do this 100 times, then calculate the probability that the dots formed by the tip of the falling marker will be within each of the different size circular regions. We finish the data collection and do our calculations, but some questions still remain unanswered when class ends. Mrs. Waters tells Mary to give me the work so that I can hold on to it for tomorrow and class ends.

As a follow-up to this strange class, I offer some additional information: Ruby's behavior was addressed by the school's "Response-to-Intervention" (RTI) team. In the following days, she was escorted from art class to chemistry class to ensure she would get there on time. The pattern of lateness disappeared for the rest of the semester. Ashley was immediately moved from this higher performing class to one of Mrs. Waters' lower performing classes for the remainder of the semester. As far as the girls suddenly vanishing from class—it was just a bit of the typical drama that high school girls can deal out. Something was said amongst them that caused all three of them to get upset with each other and they spontaneously exited the classroom. The problem was quickly resolved by the Dean of Students, but not in time for them to return to class.

But, the story of strange Day 38 doesn't end yet!

LAB SAFETY AND THE DAY WE ALMOST LOST MRS. WATERS

During the morning of April 5, 2011 (Day 38), Mrs. Waters had stopped by my office to ask my advice as to whether I thought candles would work for her upcoming "Flame Test Lab." At that moment, I was updating my chemistry notebook just in case she decided to do a notebook check in class today. She had previously hinted at this. She did notice what I was doing and smiled. I stopped my work, and told her I didn't know whether a candle would work. She tells me that her alternative is to use Sterno cans as burners—the kind used for outdoor camping stoves. We agree that it seems like we are in need of a small, preliminary experiment. She leaves to test both.

Later that day, after classes had ended, I went to check on how Mrs. Waters was doing with regard to testing both the candle and the Sterno setups. She tells me that the candles are not co-operating, but the Sterno cans seem to work pretty well. So, I proceed to help her assemble six burners: each comprised of a Sterno can and an attachable contraption outfitted with a valve for turning the gas on or off.

At one point, Mrs. Waters is having major problems screwing one of the attachments onto a can. The gas is leaking and she can't get it sealed properly—and we are both getting a good whiff of it. She is finally able to screw the attachment into a position that stops the leak, but the assemblage is still not correct. I'm having good success with mine, so I confidently ask Mrs. Waters to pass me the faulty burner. As soon as I begin working on it, the gas starts leaking heavily. It makes an extremely loud, hissing sound. The smell instantly fills the room. She runs for the windows. I immediately realize I've now jammed the contraption into a position that can't be reversed. The smell is getting worse! I don't know what to do. I panic and run for the door yelling back to Mrs. Waters "We're in big trouble now! I'm out of here. Whatever you do, don't light a match!"

Halfway through the door, I stop and realize what I am doing—I am abandoning poor Mrs. Waters, a mother and wife. I felt a surge of guilt and decide to return and give it one more try. I begin morphing into Red Adair (1915–2004), the famous oil well firefighter and rig-capping expert, and dive headfirst for the problem burner. I grip the attachment and with my shear upper body strength am able to securely seal the leak—thereby saving Mrs. Waters' life, together with the school—and perhaps avoiding the evacuation of local businesses. Hours later, the fumes have dissipated to a point where human re-habitation of the classroom is possible. At least, Mrs. Waters and I had a good laugh about the incident and my behavior; and, she got some mileage out of telling the story in the teachers' lounge!

Thursday, April 7, 2011 (Day 40) 7:15 a.m.–7:30 a.m. Today was going to be our fourth of five labs, the "Flame Test Lab." Once again I had not completed the pre-lab activity homework assignment. It was due in a few hours and I found myself sitting at my desk completing it at the last minute.

9:50 a.m., "Flame Test Lab." I arrived to class to find that the room had been reconfigured to six workstations. On the board, Mrs. Waters has written:

What will we be doing? 4/7/11

1) POD
2) Quiz on pre-lab
3) Flame test lab
4) Check H.W.
5) Finish e-prob. lab

Objectives: students will be able to

- Observe the flame color of various known metal ions
- Build a reference table of these colors
- Observe the flame color of an unknown ion and identify it

Prior to beginning the lab, we completed a POD (Problem of the Day), took a quiz, and Mrs. Waters checked our pre-lab activity homework assignment. The tables had already been rearranged for the experiment today. Six stations had been set up. We were dealing with burners, so Mrs. Waters had us clear the tables of everything except those items required for the lab. We were given clipboards to ensure that there would be no papers on the tables. Also, she reviewed the safety issues associated with today's lab and reminded everyone of the requirement to wear safety glasses. Mrs. Waters divided us into groups of three. I was put with Lindsay and Jessica G. The lab required that groups rotate from station-to-station and perform a specific procedure at each one. The experiment involved observing and recording the color of the flame that was being created. I am severely color-blind so I had to rely entirely on my lab partners for data collection. After completing the cycle of the six stations, we then had to answer several questions on a lab worksheet. Just as we began, I was suddenly called out of class by Mike Skeldon—he was looking for RIDE's graduation rate data for our school for some Brown University students who were conducting research. Within 15 minutes, I had located the information in my office for Mike and returned to chemistry class. I found Lindsay and Jessica G. struggling to answer the questions on the worksheet. They said they were confused. I was disappointed to find that they had only answered two of them. Had I been there, I'm sure we would have been further along.

Thursday, May 26, 2011 (Day 70), 9:50 a.m. (start of class) "Identification of Reaction Lab." On the board, Mrs. Waters has written:

> What will we be doing? 5/26/11
> 1) Check pre-lab homework
> 2) Do "Identification of Chemical Reaction Lab"
> 3) Label all rxn types
> 4) Check H.W.
> 5) Finish e-prob. lab
> Objectives: students will be able to
> – Distinguish between the 5 types of chemical rxn's
> – Describe through visual observation how substances react chemically in characteristic ways with other substances, with different chemical properties.

Today we performed the last of five labs for this semester. It was called the "Identification of Reaction Lab." I had completed the pre-lab activity homework assignment last night. Consistent with all four previous ones, Mrs. Waters had the room set up for today's lab prior to class. However, unlike the previous ones, this one did not require moving tables around, but did require setting up stations with the normal table positions. She checked our homework and assigned group members. Today, I was grouped with Sima, her twin sister Sarah, and Manar.

About halfway through the lab, Tina Go, the school's Admissions and Operations Manager, entered the room accompanied by students and teachers visiting from the K-8 Learning Community Charter School (Central Falls, Rhode Island). Tina sees me sitting with my fellow students and feels obligated to explain to the young visiting students and their teachers that I had stayed back many times—that's why I look like I'm older than the rest of the class.

Sima, Sarah, Manar, and I worked extremely well together, completing the simple, but rather lengthy lab procedure prior to the end of class.

SOME GENERAL OBSERVATIONS ABOUT MRS. WATERS' LABS

During the course of the 79 days I spent in Mrs. Waters' classes, I participated in all five laboratory experiments. There were definite consistencies in all of them:

1. Every lab was preceded by a pre-lab activity worksheet given as a homework assignment prior to the day of the lab. No one could participate in a lab experiment if they had not completed this.
2. Safety concerns were always addressed prior to beginning the labs. Safety glasses were provided when the experiment dictated a need for them and usage was mandatory. Special attention was also given to instructing students on the handling of hazardous materials.
3. Mrs. Waters handled the most hazardous materials herself.
4. Every lab was already set up prior to students entering the classroom. When required, tables had been repositioned into stations and all of the necessary materials had been put in place. The preparatory setup work done by Mrs. Waters made for a good start for the labs. Everything seemed well organized.
5. Lab partners were always assigned by Mrs. Waters' and lab partner assignments varied from lab-to-lab.
6. Every lab was well choreographed and there was virtually no downtime. For all practical purposes, in all five labs, the "hands-on" activities (e.g., data collection) were completed in the allotted time. There was only one exception (the "Oxygen Content in Air" lab was intentionally designed as a multi-day exercise). There were times when data analyses, write-ups, etc. were assigned as homework or had to be completed during the following class.
7. Mrs. Waters was always actively involved in continuously monitoring each lab station to ensure that students were on task. Also, she routinely posed questions to them about what they were doing thereby providing multiple opportunities for inquiry-based teaching and learning.

8. Every lab that required a significant set-up was held on the same week day—Thursday. Mrs. Waters had the first period free on that day and this allowed her to set up the room. The one exception was the "Electron Probability Lab." It had minimal set-up requirements and was conducted on a Tuesday.
9. Immediately following each lab, there was a full class discussion of results.
10. Because of the quality of the resources and instruction Mrs. Waters had developed for these labs, there was very little confusion on the part of students as to what they should be doing. In fact, I never recall hearing someone ask "What do we do next?" Even when students in their respective lab groups finished at different times, Mrs. Waters had something prepared for them to work on. In other words, students worked from bell-to-bell.
11. Mrs. Waters had built a culture in the lab such that students took it upon themselves to clean up after themselves with little or no prodding.

ASSESSING MRS. WATERS' LABS ON THE LABORATORY EXPERIENCE CONTINUUM: "COOKBOOK," "GO PLAY!" OR SOMEWHERE IN BETWEEN?

Certainly there were tasks that were of cookbook nature, but Mrs. Waters had interleaved questioning, both oral and written, throughout the lab experiences. Thus they required depth of thought well beyond merely following a cookbook recipe. And, there were clear directions and expectations with regards to student learning objectives, so this was hardly a free-for-all "Go play!" exercise. It is quite clear to me that the answer to my earlier question is a certifiable "somewhere in between." All of the lab-related activities during the semester provided multiple opportunities for inquiry.

During the several days in which I participated in Mrs. Waters' chemistry labs, I made quite a number observations in my journal on student engagement and what I perceived as teacher best practices specific to the lab activities. Out of curiosity, I consulted the *National Science Education Standards* published by National Academies of Science (NAS 1996) so as to be able to support my observations with language from a credible source. I was, after all, a novice observer. The following language is excerpted from the teaching standards I found in NAS (1996, 30-46). It aligned with my evidence-based observations in Mrs. Waters' labs:

- *Teaching Standard A.* Teachers of science plan an inquiry-based science program for their students. In doing this, they develop a framework of year-long and short-term goals for students; select science content and adapt and design curricula to meet the interests, knowledge, understanding, abilities, and experiences of students; select teaching and assessment strategies that support the development of student under- standing and nurture a community of science learners (p. 30).

- *Teaching Standard B.* Teachers of science guide and facilitate learning. In doing this, they focus and support inquiries while interacting with students; orchestrate discourse among students about scientific ideas; challenge students to accept and share responsibility for their own learning; recognize and respond to student diversity and encourage all students to participate fully in science learning; encourage and model the skills of scientific inquiry, as well as the curiosity, openness to new ideas and data, and skepticism that characterize science (p. 32).
- *Teaching Standard C.* Teachers of science engage in ongoing assessment of their teaching and of student learning. In doing this, they use multiple methods and systematically gather data about student understanding and ability; analyze assessment data to guide teaching; guide students in self-assessment; use student data, observations of teaching, and interactions with colleagues to reflect on and improve teaching practice; use student data from observations of teaching and interactions with colleagues to report student achievement and opportunities to learn to students themselves as well as to teachers, parents, policy makers and the general public (pp. 37-38).
- *Teaching Standard D.* Teachers of science design and manage learning environments that provide students with the time, space, and resources needed for learning science. In doing this, they structure the time available so that students are able to engage in extended investigations; create a setting for student work that is flexible and supportive of science inquiry; ensure a safe working environment; make the available science tools, materials, media, and technological resources accessible to students; identify and use resources outside the school; engage students in designing the learning environment (p. 43).
- *Teaching Standard E.* Teachers of science develop communities of science learners that reflect the intellectual rigor of scientific inquiry and the attitudes and social values conducive to science learning. In doing this, they display and demand respect for the diverse ideas, skills, and experiences of all students; enable students to have a significant voice in decisions about the content and context of their work and require students to take responsibility for the learning of all members of the community; nurture collaboration among students; structure and facilitate ongoing formal and informal discussion based on a shared understanding of rules of scientific discourse; model and emphasize the skills, attitudes, and values of scientific inquiry (pp. 45-46).

In chapter 8, ("Mrs. Waters' Best Classroom Practices"), I comment at considerably greater length on the many "best practices" that I observed during my 79 days in Mrs. Waters' Chemistry Block 1 class.

NOTES

1. This quote was transcribed from a 1964 lecture given by Richard P. Feynman at Cornell University and video-taped. It was included in a December 21,1993 PBS:NOVA episode entitled *The Best Mind Since Einstein,* a biography of Feynman.
2. See Brookhart (2013) for a description of various types of rubrics as well as a distinction between them and scoring schemes.
3. Depth of Knowledge (DOK) is a system for measuring the depth of understanding required to answer, explain, produce an item related to assessment, or to perform an assessment. The concept was developed on the basis of research conducted by Norman L. Webb, a senior research scientist for the Wisconsin Center of Education Research, in the late 1990's.
4. The strength of the Pearson *r* correlation of .696 is considered moderate-to-strong based on guidelines presented in Hatcher and Stepanski (1994). Approximate guidelines for describing the strength of *r*-values are found in this same work. They are: ±1.00 = perfect correlation; ±.80 = strong correlation; ±.50 = moderate correlation; ±.20 = weak correlation; and, ±.00 = no correlation.

Chapter Five

Mrs. Waters' Exit Survey

What Does the Research Say?

The winter months of 2011 were particularly unkind to Rhode Island. Most school districts had to extend their school year to make up the days missed due to snow. Beacon Charter High School being no exception, our school year was extended five school days to Tuesday, June 28, 2011 (the 180th school day). On Monday morning, the day before the last official one, I stopped by Mrs. Waters' classroom. There were no students around as she had been teaching only juniors this semester and they had been let out two weeks ago. They had started the school year two weeks earlier than the other three grades to attend an intense late summer boot camp in preparation for the October state testing in the areas of reading and mathematics (only juniors are tested). They had been promised an early dismissal (still fulfilling the 180 school day requirement) if they maintained our very high score in reading and significantly raised our dismal scores in mathematics—both of which they accomplished. Mrs. Waters was busy organizing her room in preparation for next fall—she would be teaching an anatomy class to all seniors in the first semester. I proceeded to ask my standard Monday morning question: How was your weekend? Mrs. Waters explained that she had spent the weekend in the hospital with her oldest daughter, 10-year-old Mackenzie, who had been admitted for severe abdominal pain. The first doctor to examine her suspected problems with her appendix and called for a pediatric specialist in this area. Unfortunately, as Mrs. Waters continued, only one such specialist existed at this hospital and he was not immediately available. Many hours were therefore consumed anxiously waiting. Finally, the specialist arrived, examined the child, and made his diagnosis: "Your daughter is FOP," he told her. "What's that," she asked. He responded "Full of poop!"

As soon as I heard this, I thought about some of the suggestions I had made to Mrs. Waters over the course of my 79 days in her class. For example: the design and use of study guides that align with her PowerPoints and the occasional assignment of flexible homework and flexible projects. She had willingly incorporated both of these without any evidence-based research. It was this willingness to try new strategies, or tweak her existing methods (e.g., to incorporate standard templates for her Chemical of the Day and Problem of the Day tasks), that impressed me. But would she eventually conclude that I was FOP?

Highly effective teachers are "life-long learners." They continue to seek ways to improve their pedagogy and to tweak (i.e., fine-tune) their lessons as appropriate. In my own experience teaching mathematics, I have found that requiring students to routinely write reflections on their assignments, reflections that both assess their learning experience and critique my teaching has proven invaluable. It has improved my teaching a great deal. Was an assignment boring or interesting? Too easy, too difficult, or just right? Am I moving too fast? Did I provide enough example problems (i.e., cover the material adequately) for the homework assignment? The student responses are often rich with qualitative data and provide far more insight to effective teaching and learning than does the purely quantitative analysis of their test scores. Teachers in K-12 education miss a great opportunity when they fail to collect and review student feedback in real-time. Once the memories of the classroom interaction fog over and students move on, it is too late.

I came across an interesting quote in Norman Atkins' foreword to *Driven by Data: A Practical Guide to Improve Instruction* by Paul Bambrick-Santoyo. "If there's teaching going on, but the student's aren't learning, is it really teaching?"[1] It seems to me that students can help answer this question. The recently released results of a $45 million three-year study by the Bill and Melinda Gates Foundation listed student evaluations of teachers as one of the three most useful measures of teacher effectiveness (Layton 2013).

With all of this in mind, on Day 77, two days prior to the final exam, I decided to follow through with my March 17 (Day 25) journal note (see ch. 4) and construct an *exit survey* for Mrs. Waters. It would allow her to explore students' attitudes toward some of the learning opportunities that she had implemented in her class. The survey would be attached to each final exam and each student would be required to complete it. Considering that these students had just taken the state science exam a few weeks ago; and this was followed by preparation and defense of their digital portfolios (a graduation requirement) as well as taking NWEA assessments and, considering that they were now preparing for final exams, adding another assessment into the mix was bordering on cruel and unusual punishment. I needed to keep this survey as short as possible, but at the same time, make it comprehensive enough to capture useful data for her. There were four areas of interest: (1) student attitudes toward completing the new study guides that aligned with her PowerPoints; (2) their attitudes toward the new concept of "flexible" homework assignments; (3) the top three choices (with regards to engaging) from Mrs. Waters' dozen or so distinctly different activities they had experienced during the semester-long chemistry course; and, (4) students' feelings towards possibly integrating science in their art class in the future. I worked with Mrs. Waters to develop the one-page survey that she would attach to her final exam. The students were never aware of my involvement—for all they knew, she had worked it out by herself. The nature of the questions enabled us to plan the scope of the analysis; also we did not feel the need to make the survey anonymous. The survey that was attached to Mrs. Waters' final exam is shown in figure 5.1. Each element of it is briefly discussed below:

PowerPoint study guide or no study guide? Mrs. Waters had a significant amount of her chemistry unit/lesson content available online in the form of PowerPoint presentations. Examples of presentations included: "The Atomic Structure"; "Distinguishing among Atoms"; "Electrons in Atoms"; "The Bohr Diagram"; "The Periodic Table"; and, "Ionic and Metallic Bonding." The activity with the PowerPoints entailed going online and copying the notes (i.e., word for word) from each slide into our notebooks. Each of these presentations contained 40 slides or more. Although they were rich in content and professionally presented, I wondered if students were taking the time to really develop a thorough grasp of the information. The mere act of copying the slides would be considered a Webb DOK Level One (i.e., recall) activity (see ch. 4, note 3). I found this task to be tedious and not engaging—perhaps some of my fellow students felt the same. So, I suggested to Mrs. Waters that she include a *study guide* activity—a strategy I have successfully used for years for my PowerPoints, movies, and videos. This involved working out a set of questions that aligned with the key content on her slides. This modification to an inquiry-based task would, I proposed, lead students to be more engaged in this learning activity. Higher order cognitive skills would be needed to develop answers to the questions it contained—they would not be merely copying down words. I use the term "develop" because, the questions that were created should not all be "scavenger hunt" type problems—some degree of strategic thinking should be occurring. Further, in the process of constructing this set of questions, Mrs. Waters would be encouraged to outline for herself the most important facts she wanted her students to learn. She agreed and subsequently designed study guides for all of her PowerPoint presentations. So, during the semester, students had experience with both the task she assigned initially of simply copying every word off of the slides, and then also with the revised one of completing a study guide. Hence, the survey's research question: *PowerPoint study guide or no study guide*?

Here is an example of the text from slide number 12 from Mrs. Waters' "Atomic Structure" PowerPoint that we were required to copy verbatim into our notebooks:

- Robert Millikan carried out experiments to find the quantity of charge carried by an electron.
- He used this value along with Thompson's *charge-to-mass* ratio and calculated the mass of an electron

The corresponding study guide question that was developed is:

What were the two things that Millikan used to calculate the mass of an electron?

To answer this question, students need to read and understand the content of the slide in order to derive the two things sought—i.e., the quantity of charge carried by an electron and Thompson's charge-to-mass ratio.

In contrast, a poorly written question would have been:

On which slide number is Millikan's method of determining the quantity of charge carried by an electron found?

This is what I consider a *scavenger hunt* type question. Students are just looking to match words. It's still better than merely copying notes verbatim, probably more engaging too—but arguably does not require much depth of thought.

One more PowerPoint example is from Mrs. Waters' "Distinguishing Among Atoms." Slide number 10 reads:

- The number of neutrons in an atom is the difference between the mass number and atomic number.

The corresponding question developed for this slide is:

Three isotopes of chromium are chromium-50, chromium-52, and chromium-53.

How many neutrons are in each isotope, given that chromium has an atomic number of 24?

To answer this question, students must first translate the verbal statement on slide number 10 to the algebraic equation:

Number of neutrons = mass number – atomic number

Next, students must recognize the -50, -52, and -53 in the question as the respective mass numbers and perform the following calculations:

For chromium-50: 50 – 24 = 26 neutrons;
For chromium-52: 52 – 24 = 28 neutrons;
For chromium-53: 53 – 24 = 29 neutrons

Both of these examples demonstrate that completing a Study Guide aligned to Mrs. Waters' PowerPoints required students to perform significant more work than simply copying the slides into their notebooks. So, would students seek the easy way out, i.e., prefer no study guide, or would they be engaged in their learning?

Should homework be flexible or specific? For all practical purposes, the most common practice assigning homework to students is where teachers create or choose the assignment and provide specific instructions for its completion. On at least two occasions, however, Mrs. Waters gave students in all three of her chemistry classes the opportunity to do a *flexible* homework assignment, one that allowed them to design their own assignment related to a general topic in chemistry. There were no limitations or guidelines for the work (e.g., paper, oral presentation, model) or for the length of time to be spent on the assignment. There was only a due date.

All flexible assignments involved student's orally presenting their work to the class and included a brief "question and answer" opportunity for the audience following each presentation. Evidence of student engagement was compelling. It was very clear in listening to student presentations, that students were very much interested in researching the topic of their choice (e.g., Sima's chosen topic on mood rings—something she had always wondered about), and were engaged for some period of time learning as evidenced by the posters, essays and reports, and models presented.

Examples included Sarah's tray of cupcakes decorated with the elements from the Periodic Table which she reported spending four hours on; Brittany's tray of pizza that she baked as she studied yeast in the process—another four-hour homework endeavor; and, Robert's 20-hour painting of "The Life Cycle of a Star" (see ch. 3, fig. 3.1). As shown in figure 5.1, the research question of interest posed to students on the exit survey asked them to choose between specific and flexible homework assignments. In addition to the flexible homework, Mrs. Waters also assigned flexible *projects*.

What are the students' three favorite activities in chemistry class? Over the course of the 79 days comprising the complete semester of chemistry, Mrs. Waters employed 12 distinctly different learning activities: (1) Chemical of the Day (COD); (2) PowerPoint study guide; (3) lab experiments; (4) reading assignment with double-entry journaling; (5) Study Island; (6) lab reports; (7) Problem of the Day (POD); (8) PowerPoint notes with no study guide; (9) Jeopardy game; (10) extra credit assignments with short presentations; (11) flexible homework/projects; and, (12) working with a partner on a handout (i.e., worksheet). The pertinent research question here centered on determining which three activities were the most engaging to students. It was Mrs. Waters' intent to adjust the frequency of the specific learning opportunity for future classes on the basis of evidence gleaned from this research. Each of these activities is briefly described below:

1. *Chemical of the Day (COD):* This always occurred at the beginning of class, although not every day. Mrs. Waters would write a chemical name, its molecular formulae, structural formulae, characteristics, and where it was used on the board. Students would then copy this information into their notebooks. A standard template was developed to improve their organizational skills and to provide an opportunity for thinking more deeply in completing the document.

2. *PowerPoint study guide:* As previously noted, the study guide was a set of questions strategically aligned to Mrs. Waters' PowerPoint presentations. It was introduced early in the semester as a modification to the standard task that she required of students, namely, to copy down all the content of her PowerPoint slides as a homework assignment. She would typically walk around the class the following day, checking that every student had completed the assignment. Students were required to keep these notes in their notebooks or three-ring binders.

3. *Lab experiments:* Several hands-on lab experiments were conducted during the semester—always on Thursdays, although not every Thursday. Generally, there were three or four students per lab station.

4. *Reading assignment and double-entry journaling:* On several occasions, Mrs. Waters assigned a reading (e.g., an article in a magazine, a reading from a textbook). Students were instructed to choose a specific number of quotes or statements from the reading and work out a response to each (i.e., double-entry response: a quote and a response) in their notebooks. Normally, Mrs. Waters would conclude the activity by having them discuss at least one of their entries.

Mrs. Waters' Chemistry
Exit Survey 2011

Please answer the following questions. I will use your feedback to help me improve my teaching the next time I teach this course.

1. There were two ways we handled my PowerPoint notes. The first way required you to copy my notes off the slides into your notebook. The second method is where I gave you a study guide that matched the Power Point and you had to answer specific questions. Please check the (a) method you preferred completing and (b) the one you thought you learned the most from:

   a. I preferred completing (**check only one**):

      ________ copying the notes from the slides

      ________ answering questions on the study guide

   b. I learned the most from (**check only one**):

      ________ copying the notes from the slides

      ________ answering questions on the study guide

2. Which homework or project style do you prefer: "Flexible" where you can design your own assignment related to a topic I assign or the traditional "Specific" way where I tell you exactly what you must do for the assignment.

   I preferred completing (**check only one**):

      ________ specific assignments

      ________ flexible assignments

3. Please choose your three most favorite things we did in class:
   (place #1 in your favorite, a #2 for your second favorite, and #3 for your third favorite on the line provided)

_______ Chemical of the Day	_______ Problem of the Day
_______ PowerPoint Study Guide	_______ PowerPoint Notes (no study guide)
_______ Lab experiments	_______ Jeopardy game
_______ Reading and Double-Entry Journals	_______ Extra credit assignments with short presentations
_______ Study Island	_______ Flexible homework/projects
_______ Lab reports	_______ Working with a partner on a handout.

4. Do you think your art class could be a place where you would enjoy doing an art project related to science NECAP (perhaps a 1 or 2 day assignment once each quarter)?
   YES or NO? __________

Thank you for your help   Enjoy your summer!   See you in the fall

Figure 5.1. The Exit Survey.

5. *Study Island:* this is a web-based paid-subscription instruction and assessment tool that she used for Rhode Island's NECAP science assessment preparation. This activity was generally reserved for Wednesdays—although not every Wednesday. The mobile laptop cart was reserved for the class. Students were required to create a "mistakes" document when questions were incorrectly answered and electronically email them to Mrs. Waters. This was a way of determining which topics warranted remedial work for the class as a whole or, also differentiating instruction for small groups or an individual student who might need such review.

6. *Lab reports:* Although every lab experiment had handouts (i.e., worksheets) that required students to answer questions, only one very formal lab report was required. It was a rigorous activity and was scored using Mrs. Waters' Lab Report Scoring System.

7. *Problem of the Day (POD):* As with the COD, this activity always occurred at the beginning of class—although not every day. Mrs. Waters would either write or project a problem on the board. All problems were concordant with the Rhode Island state science standards, the GSEs (Grade Expectation Standards), and provided an opportunity for preparing students for the state's NECAP science assessments which were to be administered in May. A standard template was developed to improve student organizational skills and provide an opportunity for them to think more deeply when completing the document.

8. *PowerPoint notes with no study guide:* As previously described, Mrs. Waters originally had students copy every word off of her online PowerPoint presentation slides into their notebook as a homework assignment. She would typically walk around the class the following day, checking that every student had completed the assignment.

9. *Jeopardy game:* This activity was used twice during the semester, and occurred the day before major tests. It served as a means of a test review. The game is very closely patterned after the popular television game show that has been aired since 1964. The game board is projected via a laptop and video projector (see fig. 5.2). It consists of five categories of different chemistry related topics with five opportunities per category. Each opportunity is assigned a point value from 10 points to 50 points. Just like the actual television game, behind each point value is an answer to a question and the response must be in the form of a question. Mrs. Waters developed all of the answers and questions. She typically divided the class into teams of three students and just like lab groups, she assigned membership. Each team takes turns and has exactly one minute to answer the question—the team that accumulates the most points is declared the winner. An example of an answer and corresponding correct response in the form of a question is shown in figure 5.2.

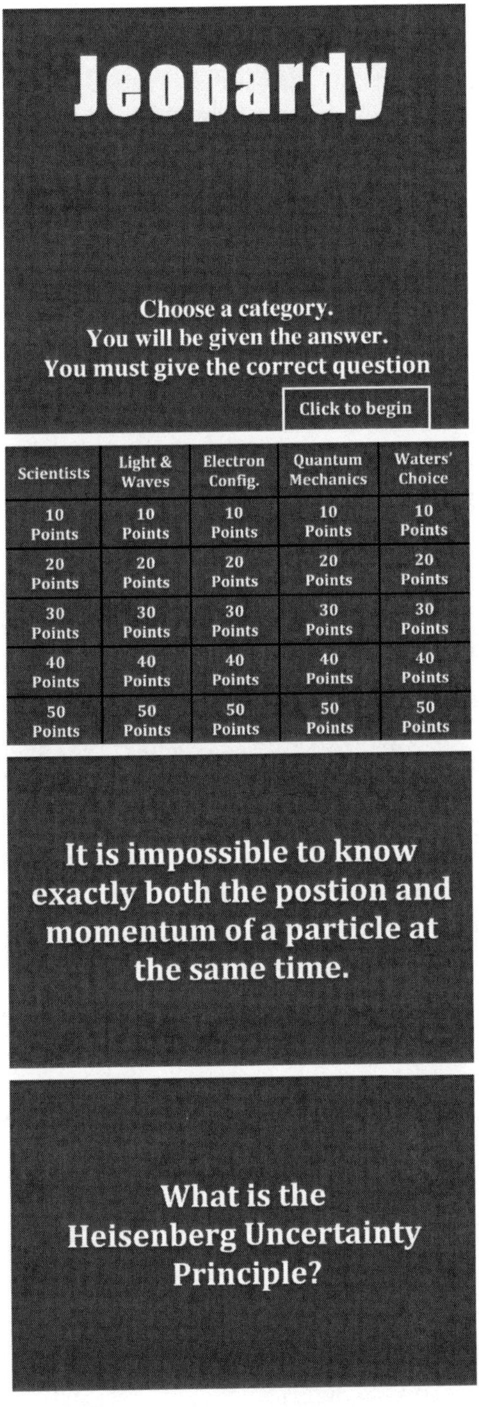

Figure 5.2. Mrs. Waters' chemistry Jeopardy game.

10. *Extra credit assignments with short presentations:* Mrs. Waters provided many opportunities for students to conduct research on various topics and make short presentations to the class for extra credit (generally five points each). They were responsible for saving their extra-credit work until the end of the semester, at which time it was applied to their lowest scores in the teacher's grade book.

11. *Flexible homework/projects:* On at least two occasions, students were given opportunities to design their own homework assignments. There were no expectations as to the time or effort to be expended on this. They were required to make a presentation to the class.

12. *Working with a partner on a handout:* Mrs. Waters had several activities during the semester that involved completing handouts (i.e., worksheets). She typically assigned each pair of students to work together.

MAJOR FINDINGS FROM THE EXIT SURVEY

There were 41 exit surveys completed by the 41 students eligible for a 100% completion rate. This group was comprised of 33 females and eight males. As concerns the arts and artistic disciplines, of these 13 were in the culinary arts, 11 in the theater arts, and 17 in the visual arts. A discussion of the major findings from the exit survey follows:

With regard to the "PowerPoint study guide or no study guide?" research question, 68% (28/41) of the students preferred completing a study guide rather than just copying notes from Mrs. Waters' slides. This left 32% (13/41) who preferred to copy notes off of the slides. One welcome finding was that 76% (31/41) of the students felt they learned more by completing the study guide, whereas only 24% (10/41) felt they learned more by only copying notes from the slides. Given these findings, it was clear that Mrs. Waters' efforts at creating study guides aligned with her PowerPoint presentations as a means of promoting student engagement in inquiry-based activity had proved fruitful.

With regard to the "Flexible or specific homework?" research question, 73% (30/41) of the students preferred being given a voice in designing their homework assignment and 27% (11/41) preferred the traditional method whereby the teacher states specific expectations for the deliverable. These findings were not surprising given the compelling evidence of student engagement collected during the presentation of flexible homework assignments.

In an effort to answer the research question: "What are the students' three favorite chemistry class activities?" I used my journal notes, class notes, and handouts from the 79 days in Mrs. Waters' Chemistry Block 1 class; her Wiki postings for her chemistry classes; and her PowerSchool online grade book to define the number of learning opportunities for each of the 12 activities across the 79 days. I next analyzed the 41 exit surveys for student responses relative to the number of first, second, and third choices of favorite activities. I recorded the frequency of the top three choices for each of the activities. Finally, to arrive at a reasonable ranking of their activities from "most favorite" to "least favorite," I devised a method of weighting the data to determine a *Weighted Ranking Score*. It consisted of awarding three points to a first choice, two points to a second, and one point for a third. Two points were given if a survey response for the three choices were made with checkmarks rather than by ordering the choices. (i.e., I assumed that there had been a tie and therefore took the average of one, two or three points = two). On six of the 41 exit surveys, students checked their three favorite activities but did not distinguish between a first, second, and third choice. Hence, each checked activity was assume a tie and awarded two points. These 12 activities are presented in table 5.1 in order of highest Weighted Ranking Score (most favorite activity) to lowest Weighted Ranking Score (least favorite activity).

As shown in table 5.1, the top three choices were: lab experiments, the Jeopardy game, and flexible homework assignments. Although lab experiments and the Jeopardy game were virtually tied as the students' favorite

Table 5.1. Ranking of Favorite Chemistry Class Activities based on Students' Responses

Ranking	*Activity*	*Number of Learning Opportunities*	*Number of Top Three Choices*	*Weighted Ranking Score*
1	Lab Experiments	5	25	58
2	Jeopardy Game	2	26	56
3	Flexible Homework/Project Assignments	3	21	40
4	Extra Credit	10	11	23
5	Chemical of the Day	12	12	22
6/7	PowerPoint with Study Guide	4	9	16
	Working with Partners on a Handout	4	9	16
8	Problem of the Day	35	4	5
9/10	Double-Entry Journal Assignments	4	2	3
	Study Island Assignments	12	2	3
11	PowerPoint no Study Guide	2	1	3
12	Lab Report (formal)	1	1	1

choice, 13 of the 41 of them chose performing lab experiments as their number one choice, whereas seven of the 41 students indicated that the Jeopardy game was their number one choice.

There is an interesting observation to be made here: Doing lab experiments was the number one favorite activity and writing the formal lab report was the least favorite.

The ranking for the PowerPoint presentations with a study guide activity was higher than that of the PowerPoint without a study guide. This further supports the earlier finding that students' significantly preferred completing a study guide and thought they learned more from this activity than they would have otherwise.

Although I took the final exam in chemistry and completed the attached exit survey as a student, I did not include my responses among the 41 student surveys. In other words, only they were used for the analyses. However, just for the record, I made the same top three choices for activities, albeit in reverse order, as did the students. My first choice was flexible homework/projects, my second, the Jeopardy game, and my third, lab experiments. As I noted in the journal kept during my fieldwork, I found myself totally engaged in the flexible assignments—even spending over an entire weekend (and about $30 out-of-pocket) on one of the assignments. With regards to my second choice, I noted that the Mrs. Waters' use of the Jeopardy game activity, strategically planned the day before a test for purposes of review and used on two occasions, saved me from low and potentially embarrassing scores on both tests. There were several questions on the game that I couldn't answer. However, hearing the correct answer and explanation as we played the game filled in the gaps in my knowledge base. It was extremely impressive to watch the interaction and hear the discourse amongst the students as they played the game correctly and answered some fairly difficult questions. It was clear that highly effective teaching was going on in Mrs. Waters' class. As for my third choice, I too found the several lab experiments engaging and wished we had done more of them.

The basis for the fourth research question, "Did students see an opportunity to integrate science in their art class? was prompted by the revelation of a possible trend observed in student work that was generated by the flexible homework opportunities. Examples included Sarah, a culinary student, baking periodic table cupcakes; Brittany, another culinary student, studying yeast by baking a pizza; Robert, a visual arts student, spending twenty hours on homework that involved painting the life cycle of a star; Siraj, another visual arts student, painting and labeling the parts of the human eye. There were many more examples of Beacon's student-artists who connected their art with science.

The underlying issue here was that Mrs. Waters had to plan her curriculum based on a limited number of instructional hours as these were dictated by Beacon's unique emphasis on the arts where course scheduling was concerned. All students have art class every day for all four years. This translates into twice as many hours of art than math, English, and science, since, up till now, these academic courses have been only one semester long. The disproportionate amount of instructional time between the arts and academics is problematic when one considers that English, science and math are the content areas assessed by the state on the NECAP. To help increase ALT (academic learning time) with regards to science classes, one might integrate science into the art classes on a very limited basis—perhaps by the addition of interdisciplinary assignments of a duration of one or two days per academic quarter. The results of the exit survey question found that 61% (25/41) felt they would enjoy doing a NECAP science related project in their art class while 39% (16/41) responded that they would not.

I checked to see if there were any significant differences of opinion across the three different types of arts disciplines taught at Beacon.

The numbers of students answering "yes" to the question about doing some science in art classes was 62% (8/13) of the culinary arts students, 55% (6/11) of the theater arts students, and 65% (11/17) of the visual arts students. Thus, a majority in each of the three areas agreed that they would enjoy doing science in their art class.

I also considered the relation between students' final chemistry class grade and their response to the science in art class question. The average grade for those responding "yes" was 81, for those responding "no" was 78. This is not statistically significant.

Beyond the research questions of the exit survey, a few additional analyses were carried out: I investigated the average final grades for chemistry of the 41 students across the three art disciplines and found no significant statistical differences: 82.9 for culinary arts, 77.9 for theater arts, and 79.4 for visual arts.

With regard to gender and final grades for chemistry, "trend-level" significance ($.05 \leq p \leq .10$) was found between females' ($n = 33$) average final grade of 81.8 and males' ($n = 8$) average grade of 73.0.

A FEW CLOSING REMARKS

Although this exit survey and the subsequent analyses are exploratory in nature, they serve as an example of how student-generated feedback can aid in improving instructional methods and curriculum decisions. Mrs. Waters may, for example, want to increase the frequency of lab experiments, the Jeopardy game, and flexible homework/project assignments given that students reported these as their three most engaging activities. She may also want to explore why the formal lab report was the least favorite. There was a compelling finding that students preferred completing a study guide for the PowerPoint presentations to the alternative of simply copying notes from the slides into notebooks. Students believed they learned more by completing the study guide. This would support the conclusion that study guides should be created for all PowerPoint presentations and that they can promote student engagement.

As concerns the difference in the average final chemistry grades between females and males, this was a welcome finding. There is a need to attract more women into STEM (Science, Technology, Engineering, and Mathematics) fields. Are there similar gender-specific patterns in other science content areas?

Also, this type of evidence-based research may help in *response to intervention* (RTI) decisions and provide guidance toward "best teaching" practices with regard to the new Rhode Island Model Educator Evaluation System.

NOTES

1. In Norman Atkins' foreword to "Driven by Data: A Practical Guide to Improve Instruction" (2010) by Paul Bambrick-Santoyo, Atkins credits Bambrick with the "schooling equivalent of the fabled tree-forest conundrum" (Bambrick-Santoyo 2010, xiv).

Chapter Six

The State Science Exam

I do not think that there is any other quality so essential to success of any kind as the quality of perseverance. It overcomes almost everything, even nature.

—John D. Rockefeller[1]

Occasionally, I make the 50-minute trip to the Foxwoods Resort Casino in Mashantucket, Connecticut where I play the slot machines. I have a system. I walk up and down the aisles until I hear a machine calling out "John, play me!" I have to walk past them a few times to positively confirm which one is speaking to me.

Once I've confirmed this, I insert my Foxwoods "Dream Card" into the flashing slot on the machine to earn points on my account. They are extremely important—something I will explain shortly. Unfortunately, I get a message: "Please reinsert card." I repeat the process twenty times and get the same message an equal number of times. This is exasperating! I need those points! Meanwhile, the machine continues with its irresistible call "John, play me." Finally, I have no choice but to get up and find another one. But no other machine is calling out to me, so I have to play one that is less desirable. Alas, I find a machine down the aisle that does read my card and I insert a twenty-dollar bill.

Suddenly, bells and whistles go off. High above, fireworks spell out the word JACKPOT!

But no, it's not for me. It's the person who sat at the machine I have just left. The only sound coming from my machine is the loud deep sucking noise as it swallows my twenty-dollar bill!

About those Dream Card points—after I've lost all my money, it's the points that pay for a sandwich on the way out.

The many years of this pattern of "lack of success" have caused me to continually fight a mental state often referred to as *learned helplessness*. It seems that no matter how much I try to succeed at slots, I will still walk out with a deflated wallet, a sandwich in my stomach, and a Dream Card containing no points.

The theory of learned helplessness was first proposed by J. Bruce Overmier and fellow researcher Martin E. P. Seligman in their 1967 article entitled "Effects of Inescapable Shock upon Subsequent Escape and Avoidance Responding" published in the *Journal of Comparative and Physiological Psychology*.

Briefly stated, it claims that an organism that continuously fails at escaping an adverse condition will eventually give up trying and learn to live with it. This theory was developed from Pavlovian conditioning-type laboratory animal studies (Pavlov [1927] 1960). They involved using a two-compartment shuttlebox with a low barrier partition between the two sides. In one test group, a dog was placed on one side and given mild electrical shocks or subjected to high pitch tones. He learned that if he jumped over the barrier to the other side, he would escape the electric shock or noise. In another test group, the dog was shocked or subjected to the high-pitched tones on both sides of the fence, so he learned that it was useless to jump over the fence. Subsequent to this, he would merely lie down when shocked or subjected to noise, hence the phrase "learned helplessness."

However, the original theory was found to be inadequate. Some subjects never gave up despite the adversity they faced. Maybe my continuous ventures to Foxwoods are an example of such perseverance!

Subsequent to the observation that not all subjects gave up, a reformulated theory was worked out in a special issue of the *Journal of Abnormal Psychology* in February, 1978. The history of learned helplessness is illustrated in Peterson, Maier, and Seligman (1993, 303).

Learned helplessness is a phenomenon that appears in many areas of life, including education. My first experience with this occurred in my first year of teaching in a Providence, Rhode Island high school when I proctored the state's math assessment. After reading the test instructions, I signaled to the class to begin. Just about every student out of the 25 put their heads down on their desk. Apparently, they had become complacent from previous years of standardized test taking and were operating on the assumption that no matter how hard they would try, they would still fail. Although I am unaware of any formal research on this, I believe that the emphasis placed on high stakes testing by the *No Child Left Behind Act of 2001* (NCLB) has contributed to the development of such learned helplessness on the part of many students. In fact, and possibly as a direct result of this legislation, many children have been left behind.

Many parents also develop learned helplessness when their child continually brings home low or failing grades. They often state: "I wasn't good in math either." Sadly, in some families, particularly in urban areas, when this negative perspective is forthcoming from adults, it spans several generations.

In recent years, great emphasis has been placed on scores on state assessments, particularly in the content areas of mathematics, reading, and science. Schools are weighed against each other when scores are published in the media; also, student scores are used for teacher evaluations and recertification, and parents purchase homes in school districts based on school performance, etc. In some cases, consistently lower performing schools receive mandated restructuring interventions or are even closed. The accountability issue carries particularly heavy weight with charter schools. So, state tests must be taken seriously. A room of test takers can't simply put their heads down on their desks because of learned helplessness or lack of buy-in, never mind lack of knowledge. Student engagement is crucial to any assessment process if it is to accurately reflect student ability.

DIGGING OURSELVES OUT OF A HOLE

Since the reinstatement of its charter in 2005, Beacon Charter High School has focused on developing a rigorous academic curriculum in an effort to enhance student achievement on the Rhode Island state assessments, the NECAP (New England Common Assessment Program). The NECAP was created in response to the federal mandates of NCLB. It was a joint effort between the New England states of New Hampshire, Rhode Island, and Vermont. At the public high school level, students are tested in the eleventh grade in the content areas of mathematics, reading, writing and science. In the near future, the NECAP (except science) will be replaced by a new national assessment aligned to national standards known as the Common Core State Standards (CCSS)[2].

It is fair to say that Beacon's performance in all of the heretofore assessed NECAP content areas has at some point, been amongst the lowest in the state—or certainly below state averages. First of all, our writing scores were very low, so to remedy this, we subsequently implemented school-wide interventions. Then it was our reading scores that needed significant school-wide attention, and last year, it was the mathematics scores. The faculty has put in extra work during summer vacations and after school in the course of the school year. In all three of these areas, i.e., reading, mathematics and writing, Beacon has risen to become one of the top high schools in the state, albeit with a slight dip in the math performance in Fall 2012. In fact, Beacon earned a U.S. News Best High School Bronze ranking in 2009 and more recently in 2011.

So, now it is time to focus on science. We were 5% "proficient or above" in testing year 2008 as compared to 16% for the state as a whole. In both of the 2009 and 2010 testing years, we were 14% "proficient or above" as compared to the state's 19% and 21% respectively. Even the state numbers on the science assessments, as measured by NECAP, are quite low.

So, for this school year, Mrs. Waters has increased the opportunities in her eleventh grade chemistry classes that directly prepare students for the NECAP science assessment in the spring. Early on in the semester, I asked her about how she planned to prepare her eleventh grade students for the NECAP. She told me that she does not believe in teaching to the test. In her view, a strong inquiry-based curriculum aligned with the state's science standards, the GSEs (Grade Span Expectations) is fundamental to student achievement on the assessments. However, embedded in her curriculum for eleventh grade chemistry classes are opportunities for review of ninth and tenth grade science content as well as of the new content being introduced in her chemistry classes. These include:

- A POD (Problem of the Day) aligned to NECAP practice problems. This activity occurs at the very beginning of class.
- Every Wednesday, the laptop cart, which is equipped with a complete classroom set of computer laptops, is brought down into the classroom so that students can access an online web-based tutorial program called "Study Island." This has been a very useful tool in preparation for the NECAP mathematics assessment and the program includes a science curriculum aligned to the standards.
- Mrs. Waters assigns "flexible" homework and projects which are differentiated with respect to individual student weaknesses in the NECAP science domains of *Earth & Space*, *Life Science*, and *Physical Science.* These are based on the student performance on the NWEA science assessments that we give to all Beacon students twice a year.
- Mrs. Waters has two different lessons which are designed to practice for the NECAP science "Inquiry Task" portion of the state's assessment.
- Science vocabulary words and definitions are distributed and practiced in non-science classes.
- Mrs. Waters has a series of engaging science cartoon books that align with the NECAP assessed content. She keeps these in her science room and rou-

tinely assigns short readings from them together with response tasks (e.g., "double entry" journaling).

Over the last few years, we have developed a model for preparing for and administering the state assessments, one that seems to be working well. Interestingly enough, the model is based on addressing the five tenets of educating the "whole child" as developed by ASCD (Association for Supervision and Curriculum Design)[3]: *Healthy*, *Safe*, *Engaged*, *Supported*, and *Challenged.*

With regard to the first on testing days, prior to actual testing, students are provided activities for relaxation such as taking a walk, playing games, or sitting in a quiet zone. They are also given healthy food and drink prior to and during the testing sessions.

As for the second tenet, on testing days, test takers are segregated from the rest of the school and placed in an environment that promotes safety and specifically makes the test environment comfortable. Special attention is paid to spacing and seating, room temperature, proper lighting, and minimization of noise and distractions. Also, two adults are assigned to each room.

Unlike my first year of teaching in Providence and first year of proctoring the state exam, by contrast to that, Beacon's students are "engaged" and try hard on these tests—learned helplessness is not a problem. Perhaps the difference in attitudes is simply due to Beacon being a charter school and charter schools are schools of choice—students choose to come to Beacon. However, I suspect it's much more than this. For example, Beacon has a student senate that provides a student voice in various school activities. There is a recycling club, whereby students take an active role in respecting the community. Students are able to be members of the schools SIT (School Improvement Team) that also includes administrators, teachers, and parents. We have even allowed a small group of students to participate in the interviewing of prospective new teachers. At Beacon, teachers hire teachers and the principal does not have a vote. After the Beacon faculty has selected a few candidates as finalists for a vacancy, we often have them come back for a demonstration lesson. In this case, we select students to act in the role of a class for the candidate. Afterwards, we listen to and consider the students opinions as to the candidate's performance—student feedback is weighed heavily.

Common to all of these reasons is that of respect and ownership—none of which were present at Fogarty. Beacon's students feel they own a piece of the decision making process about the operation of Beacon and thereby respected as stakeholders. With this autonomy, comes the awareness of the importance of academic performance. They are aware of the significance of the state assessments both for them personally and for their school.

With regard to being "supported" in this process, multiple opportunities are afforded to them to prepare well for the state assessments. Perhaps most importantly, they know that their teachers care about them and want them to succeed.

And, they feel that they are "challenged" by the science curriculum because of its rigor (e.g., inquiry-based instruction, homework routinely assigned, papers, projects and presentations required). Recall, as previously mentioned, that dropouts have stated they wished they had been challenged more (Bridgeland, DiIulio, and Morison 2006).

Beacon students who have taken the NECAP science assessments in the past report that they had been adequately prepared.

THE COUNTDOWN

Friday, May 13, 2011 (Day 61), 5:01 a.m. It's Friday the 13th and there are only two chemistry classes left before NECAP testing. I'm half-asleep when the sound of an incoming text message jolts me awake. It is from Mrs. Waters:

> John, I need you to help me out today…I haven't even gone to bed yet. I am so sick, fever, chills, vomiting. I have been trying to fight it but I can't even stand up. Chris has to take me to the emergency room. Can you please have Block 1 finish the Periodic Table thing? I'll send you the Problem of the Day. Block 2 and 3 need to finish going over the "Location Inquiry Task." Maybe Shayna can help. Everything is in my mailbox. I am beating myself up about this!! I don't think Mike is in, that's why I'm texting you.

5:03 a.m. I respond: "No problem. I'll take care of everything—relax this weekend."

5:05 a.m. Mrs. Waters texts back: "Thanks so much, I really appreciate it."

6:45 a.m. I arrive at school. Mike Skeldon (Academic Dean and Principal in Residence) is in the Teacher's Room, i.e., lounge working on coverage for Mrs. Waters' classes. He has me taking care of DEAR (Drop Everything and Read) and covering Mrs. Waters' Chemistry Block 1 class. So, today I'll have to wear my teacher's hat.

7:30 a.m. I've been checking my email every five minutes but have not yet received the POD (Problem of the Day) from Mrs. Waters. I decide to run copies of three practice NECAP problems for the class in case it does not arrive.

8:00 a.m.–9:45 a.m. I take care of DEAR and our Block 1 class. Everything went smoothly. I read the instructions that Mrs. Waters had left in her mailbox for the class assignment and passed out the three NECAP practice problems to be completed after her assignment was done. We had time to review the answers before the end of class without any significant issues.

3:02 p.m. I receive a text from Mrs. Waters: "Just got out of the ER. I was extremely dehydrated. Had to have three bags of fluid. Very bad case of food poisoning. I feel like I'm letting everyone down for NECAP."

3:05 p.m. I text back: "You're not letting anyone down. In addition to your instructions, I gave everyone three PODs. Call me when you feel better!"

5:30 p.m. I left school today knowing I had a lot of work to do over the weekend. I need to put on my Data Analyst hat and figure out the seating plan for Mrs. Waters for the NECAP science testing next week. I will use the same method I used for the fall assessments and arrange students from "highest" to "lowest" with respect to their achievement on the NWEA science assessment. Then I will fill the designated classrooms on the basis of this achievement. My reasons for this will be outlined later in this chapter.

I also need to work on my "flexible" project assignment for chemistry. It is to be presented on Monday.

Monday, May 16, 2011 (Day 62). Today is the day before NECAP testing, and it's not off to a good start. I had gone to bed at 2:00 a.m. after working on my "flexible" chemistry project to no avail—a major technical failure occurred. I arrive at school around 7:30 a.m.—about an hour later than usual. As I was signing in at the Main Office, Mrs. Waters walked in. She told me more about her visit to the ER last Friday. Her food poisoning was from a tuna fish sandwich she had eaten at a local Chelos the night before. She asks what I thought of the room changes for the NECAP science testing. Her question catches me by surprise. "Did you see Mike's email?" she asks. I answer in the negative. She tells me she checked her email over the weekend and found Mike's schedule for this week. It reflected a change in plans for NECAP Science testing—we will be using the Theatre room and Mr. Boland's English classroom, both on the third floor, instead of the fourth floor that was the location for all previous NECAP testing.

Mrs. Waters seemed upset about this last minute news given that testing will begin tomorrow. I was a little shocked myself because Principal Pilkington often talked about the NECAP testing model that had been created at Beacon. He believed that it was a major factor in the school's recent significant increases in the NECAP Reading and Writing assessments. Any tampering with this success would seem foolhardy.

"High stakes" testing, like the state's assessments, can be a stressful experience for teachers of science, math, and reading. This is especially true in small schools like Beacon where there are only one or two for each content area. It becomes very easy to link teacher with student test results. In fact, teachers in the NECAP assessed content areas often feel they have a target on their back. To defuse this pressure, administrators will often tell the whole faculty that everyone in the building owns the results—good or bad. These are nice talking points, often spoken with good intentions. However, the bottom line is that science and math are not taught in art and gym.

If the science scores go down, Mrs. Waters is going to feel, that she let down not only her students, but also the school. It will bother her for a whole year until the next test is administered. This same grinding, year-to-year mental cycle holds for most math teachers and also for English teachers. It's definitely high stakes—this new era of increased accountability and *data-driven* education is increasingly linking student performance on state tests to educator employment decisions (e.g., teacher retention, certification, merit pay).

So, I can appreciate Mrs. Water's nervousness about this. The test will begin in about 24 hours. She knows that she has done her level best with the time she has been given to prepare her eleventh-graders, but she will need every student to try their hardest.

Mrs. Waters and I agree to meet later on today to work on the seating arrangements for the testing rooms using my NWEA data. I tell her that I will go to Mr. Boland's room to determine the appropriate number of students to be seated there.

Back in my office, I check my email and find the message that Mrs. Waters referred to. Mike Skeldon had indeed sent it to the entire faculty. Attached to Mike's email is an EXCEL table that shows the schedule for the NECAP testing and the room assignments. It is clear that it will be done on the third floor using the Theatre Room (302) and Mr. Boland's Room (303) next door.

Last year, I had come up with a system for the seating arrangement for the NECAP math exams. It seemed to work well and I wanted to follow the same methodology for the NECAP science exams this year, I would propose that we stock the smaller room (i.e., now Mr. Boland's) with the students who were most likely to score the highest on the tests. This was based on my hypotheses that higher performing students are more likely to be able to work independently; more likely to try harder; more likely to understand instructions and less likely to ask questions; and they are also more likely to stay focused and less likely to give up in frustration. Further, higher performing students are more likely to finish earlier than lower performing ones. This can, however, add to the stress of lower performing students. They feel academically inferior because they are not close to finishing. They may quit prematurely so they can leave early and give the rest of the test takers in the room the impression that they have breezed through the test with ease. Thus, I prefer to group them by their latest NWEA test scores in testing rooms so as to minimize such distractions. Of course, this all has to be done in such a way that students do not perceive there is a "smart" room and a "not-so-smart" one. Fortunately, this has not been a problem in previous testing for two reasons. First, one room is substantially larger than the other, so there is a limitation of 18 students for the smaller room–students understand this. Secondly, students generally do not know each other's NWEA scores and certainly don't know my strategy for seating them.

In either testing space, they are unlikely to feel confined. The third floor testing rooms are actually larger than the respective fourth floor testing ones used in the past. There are no noticeable problems with lighting, temperature, or noise in either of the third floor testing rooms except for the annoying trickle of water from Boland's fish tank pump. "Fish sticks for lunch?!"

Our NECAP science performance results for the last two years have been a stagnant and remarkably low 14%. Considering the significant attention given to NECAP preparation by Mrs. Waters and my data for this year's students, I have to believe the results will be better overall.

1:45 p.m. I count seats in Room 303 for NECAP. Mr. Boland has it set up with nine rectangular tables and two students per tables—exactly what I need. We will put the top 18 students in this room.

1:55 p.m. Mrs. Waters stops by my office to ask about the seating plan for tomorrow's testing. I tell her that Boland's room is all set and hand her the list of 43 test takers with their names listed in order by NWEA scores.

Mrs. Waters tells me there's some controversy over the room changes and confusion as to who is responsible for the food and drinks for PRECAP. And, there is no Wii room (Nintendo video game system)—just the "chill" and "walk" options. I wanted to hear more, but she didn't have time to explain. She was off to a meeting on NECAP with Robin Murphy (Dean of Students).

2:00 p.m. Meeting on NECAP. Mrs. Waters gave Robin the list of student names for Boland's room and explained how we generated it. Robin informs her that Ruby (pseudonym) will be suspended for one day because of dropping the "F" bomb at the time the Learning Community Charter School (Pawtucket, Rhode Island) students and teachers were visiting. The timing of her suspension couldn't be any better given that she can be such a distraction in the classroom. She will have to take the exam on make-up day.

Mrs. Waters asks Robin about the room changes. Robin says there are ten seniors who are failing their Capstone[4] project and may not graduate. Uriah Donnelly (Beacon's film teacher) needs to work with them on the special editing computers in the Film room—the fourth-floor room that was used for NECAP testing in previous years.

2:30 p.m. Mrs. Waters stops by before she leaves for the day and fills me in on the reasons for the room changes. She tells me that she's disappointed and upset about the changes, but the administration has made a decision and she doesn't feel she should fight it.

3:15 p.m. Dr. Pilkington stops by my office to ask if I had heard how the NECAP meeting went. My answer: "Fine, I guess, except that Mrs. Waters mentioned earlier that there was some controversy involving the sudden room changes and the changes to PRECAP."

Dr. Pilkington's demeanor suddenly changes to extreme anger—apparently he had made his wishes known to both Mike and Uriah about having the NECAP testing remain on the fourth floor. It was a formula that was known to work. He says Uriah walked out of the meeting. I explain that both Mrs. Waters and I were out of the loop on this decision and that we only found out about the schedule change by reading Mike's email over the weekend—the decision must have been made on Friday when she was out sick.

Dr. P was extremely angry and wanted to know if Mrs. Waters had made her feelings known at the NECAP meeting. I couldn't answer on her behalf. However, I did explain that she was disappointed and upset about it. Robin had, however, given her a very logical reason for the change: ten students were failing Capstone and Donnelly needed the room to work with them. Dr. P comments that they could have moved the computers. He calls in Pat Hawkridge, the school's Artistic Director and expressed his feelings to her. She agrees with Dr. P.

I would add here that Mrs. Waters did not want to get involved because she felt that Mike had made a decision. She didn't want to get him in trouble.

Still, Dr. P. left the school very angry.

7:00 p.m. Mrs. Waters calls me and we have a long conversation. She restates her position on this: She didn't fight the decision because she felt that, as a teacher, she should not argue with an administrator.

Tuesday, May 17, 2011 (Day 63), 5:15 a.m. Today is the first of three sessions for the NECAP science testing. After a restless night—too often laced with moments of complete consciousness—I awoke with the following thought: we will never know for sure what the effect of the room changes may have on the NECAP scores. What we do know is that the pre-test activities/test environment model used in the Fall NECAP test administration of math, reading, and writing over the last two years was part of an equation that produced significant positive gains in student achievement. From my perspective as the school's Data Analyst, I knew that this room change would complicate things. Statistical inferences on student performance could be compromised. The best I could perhaps do would be to construct a post-testing survey that would assess students' feelings towards the change in venue. I made a note in my journal to remind myself to do this.

7:15 a.m. I arrive at school and stop in the Main Office to sign in and check my mailbox. Nobody is around. Between the Main Office door and the teachers' mailbox, I walk past a couple of empty chairs—they look very inviting at this point. I plop myself down into one of them and decide to rest a bit before making the journey up to the fourth floor via the elevator. Just as I am catching my breath, Dr. Pilkington arrives. Looking a bit disheveled, he sits in the empty chair next to me. He tells me that he's only had four hours of sleep and did not talk to his wife last night because of the room change problem; also she is dealing with her own problems. Dr. P's wife is a teacher and a union building representative in a Providence middle school. The Providence Teachers Union (PTU) is currently dealing with the firing of 1900+ teachers (i.e., basically everyone in the city) and also with the problem of the 400+ who would not be hired back. This is an item that has made the national news.

I mention my thoughts about the change in venue and the fact that it might complicate future analyses of the test results. But I limited my statement to less than 20 words because I saw the metaphorical steam coming out of his ears. He quickly got up and went over to the receptionist's desk where he wrote a note to Mrs. Waters, Robin Murphy (Dean of Students) and myself stating he wanted to see the three of us in Robin's office as soon as everyone arrived.

7:30 a.m. Mrs. Waters arrives and says she didn't sleep well and had a dream about only doing two of the three required sessions of the NECAP. I hand her Dr. P's note.

7:45 a.m. Robin Murphy arrives and Mrs. Waters and I pass the note to her. We head down the hallway together to Robin's office on the first floor. She is sitting behind her desk. Dr. P. is on my left and Mrs. Waters on my right. The three of us are facing Robin.

Dr. P starts off by repeating the story he told me, now for the benefit of the other two who were not present earlier, about only having four hours of sleep and not talking to his wife because she was having her own problems. He tells us that he made it clear to Mike and Uriah that he wanted NECAP on the fourth floor because we had established a formula that worked. However, Dr. P explains that when he told Uriah: "When the state starts publishing Capstone results in the paper, you can have the fourth floor," he had walked out.

Mrs. Waters then speaks up. As far as she was concerned, an administrator had made a decision and she felt she couldn't and shouldn't fight it.

Now it is Robin's turn. She says that she didn't care on which floor the NECAP was held, but that Mike made the decision and that Mike is the Principal in Residence. Like Mrs. Waters, Robin respected his right to make the call.

Now, I have known Mike for over five years and I understand that he does not make decisions without first reasoning things through. He definitely had a good reason for changing the venue. So, it now seemed like it was time for me to chime in. I suggested to Dr. P that he should not undermine Mike's decision by demanding that the fourth floor be used for this. Looking at the "big picture" over the longer term, forcing Mike to reverse his position on this would be harmful to Dr. P's mentoring relationship with him. He might be left with the feeling that any decision he makes in the future could be overridden by Dr. P.

The meeting adjourned. It was clear that Dr. P was still not happy, but at least he had calmed down and had a chance to vent. Mrs. Waters and I now have to focus on our responsibilities with regard to administering the test. As lead science teacher, she is the principal test administrator and I am merely one of the proctors.

8:10 a.m. The eleventh-graders have been instructed to report to the cafeteria. I enter it to find all of the juniors seated there. Pete Boland, the school's English teacher, is standing on a chair near the center of the room speaking to the group. He was blessed with a great speaking voice—I think he told me once that he had some radio broadcasting experience. He's giving the juniors one of his patented motivational speeches to get them ready for the first day of the state's NECAP science tests. "Everyone take out a piece of paper and something to write with," he shouts. "Okay, now I want each of you to write down five reasons why you are intelligent." He then asks for volunteers to give an example of what they wrote. One by one, students raise their hands and respond when called upon. It's pretty impressive to see all 43 of these students gathered in the cafeteria and engaged in Mr. Boland's motivational exercise.

8:30 a.m.–9:25 a.m. It's time for PRECAP. Students were given a sign-up sheet yesterday and asked to select one of three pre-test activities. This has become known as PRECAP (I think Carrie Appel, Lead English teacher and future Academic Dean, may have first coined the term last year). On the sheet, the activities were described as:

- "Walk on Waters"—students take a walk outside through the Beacon neighborhood with Mrs. Waters (they must submit a permission slip signed by a parent or guardian).
- "Fly on the Wall"—It's quiet time to relax, read and listen to music with Miss Wall (the school's humanities teacher) in Room 208.
- "Muzzle-free Mowry"—Chat and chill with Mrs. Mowry (Spanish teacher) in the Café´.
- "Brain Food and Bagels with Butler" (9:25–9:45 a.m.)—Fill your bellies and energize your minds with fruits and grains in the Café.

9:25 a.m. Brain food is served. All test takers have gathered in the cafeteria. Several mostly healthy food choices have been provided for them: bagels, both plain and cinnamon; muffins, chocolate chip, banana, and blueberry, with butter and cream cheese; yogurt, strawberry and banana; assorted cereal bars; trail mix; apples and oranges; orange, grape, and apple juice; and, cold water.

Bottles of water and cereal bars have been packed into plastic milk crates to be taken up to the testing rooms in case test takers need such refreshments during the periods of the testing. Also, there are bags of peppermints in the crates.

As reported in a *Washington Post* article (Tuesday, March 20, 2007) entitled "The Power of Peppermint is Put to the Test," author Lori Arafani cites the following research: "In the 1990s, researchers at University of Cincinnati found that a whiff of peppermint or muguet[5], a scent similar to lily of the valley, helped test subjects concentrate and do better on tasks that required sustained concentration. Joel Warm, a professor of psychology who conducted this research with his late colleague William Dember, said there is more than a bit of truth in the peppermint theory: "Not only do you get an improvement [in focus] with peppermint, you get a change in response that affects alertness in target detection." (see Warm, Dember, and Parasuraman 1991). I don't know if mints work or not, but we did serve them during the fall NECAP assessments and our scores went up in all categories! Maybe we should spike the water too!

9:40 a.m. As we finish filling our students bellies, I pull out my assignment sheet for the room and proceed to call out the names of those who will be taking the test in Mrs. Mowry's room and those who will be in the Theatre room.

9:45 a.m. Students report to their assigned rooms. Carrie Appel and Mrs. Waters will proctor the exam in the Theatre room while Shayna Fox (Reading Specialist) and I are in Mr. Boland's room.

9:45–10:00 a.m. Attendance is taken and test materials are passed out per the NECAP protocols. I read the prescribed instructions to the 18 test takers who are seated in front of me, and then the last sentence: "You may begin."

10:00 a.m. Work begins. The estimated time for completion for this session is 45 minutes. However, students are allowed to continue working beyond this as long as they are being productive. Everything is going well. The only issue is that annoying and distracting gurgling noise coming from Boland's big fish tank. I unplug the pump and the room suddenly relaxes into silence. Everyone appears to be working hard. Nobody is giving up. I experience flashbacks to the images of the "boy under the coat" (ch. 3, "The Aesthetics of Learning"); and, my first experience of proctoring the state exam 10 years ago in Providence, the one in which every student put their heads down on their desks. Mrs. Waters is floating between both the Theatre room and this one to check that everything is going well.

11:30 a.m. The last few students finish and leave.

Wednesday, May 18, 2011 (Day 64), 9:40 a.m. Today is the second of three sessions for the NECAP science test. I retrieve the secured test materials from Mrs. Waters, sign for them, and then proceed to Boland's room to proctor it.

9:55 a.m. I read the instructions to the students as I had done for the first session.

10:00 a.m. Work begins. The estimated time for completion of this session is 45 minutes as the first one. And again, students are allowed to continue working as long as they are being productive.

10:15 a.m. Shayna Fox takes charge while I go across the hall to the server room. I need to add the names of two students to the list for NWEA testing. It is going on simultaneously in the computer lab upstairs. While I'm working on the server, Dr. P stops by. He wants to know if I want to hear something that will really get my blood boiling. Of course! When he puts it that way, how can I refuse? He proceeds to tell me that there is only one student up in the Film room—the room that was the focus of the room change controversy. Recall that we have changed floors, relocating 43 test-takers to the third floor, so that 10 students could work on their Capstone films in the Film room. Obviously, he is still angry over this change. I tell him that there really haven't been any problems with the new venue. He leaves (probably not happy with my answer) and I finish my work and return to Boland's room to assist Shayna.

Everything is going well until somebody starts banging on the drum set up in the Film room (above Boland's room). The noise is loud enough to send ripples across the surface of the water in his fish tank! The Film room is also used for the afterschool band that meets every Friday afternoon. We have several musical instruments stored there. Unfortunately, this is not the time for someone to be banging the drums and interrupting the work of the students who are taking this test. I'm angry and dive for the phone to call Uriah Donnelly. He should be up in the Film room, but there's no answer. By coincidence, he is just entering our room. He hears the banging and immediately reverses direction and runs back upstairs. Needless to say, it suddenly stops.

11:30 a.m. The last few students finish and leave.

Thursday, May 19, 2011 (Day 65). Today is the last of three sessions for the NECAP science test—the final day. Students are asked to complete a "Scientific Inquiry Task" in which they hypothesize, plan and critique investigations, analyze data, and develop explanations. Typically, there's a short article or story with data and several questions to which they are expected to respond. Mrs. Waters had done targeted practice on these types of problems in her three chemistry classes, so, hopefully, her students will do well.

8:00 a.m.–8:30 a.m. All of the junior advisors are asked to report to the cafeteria before PRECAP activities today. Pete Boland stands on a chair at one end of the room and proceeds to hand out certificates to those students who achieved "honors" or "high honors" for the grading period that just ended. All of the eleventh-graders have had Pete as their English teacher at some point and he has made up his own nicknames for each of them–decidedly not one of the "best practices" listed in chapter 8, "Mrs. Waters' Best Classroom Practices."

However, over the years, Pete has used this "nicknaming" strategy as his means of getting students to feel they are getting special attention. The names are never derogatory (at least from Pete's perspective and intentions) but are almost always related to a student's first or last name. For example, Jessica G. in Mrs. Waters' Chemistry Block 1 class (whose last name begins with "G") has the Boland nickname "Lady GooGoo" named after current singer/songwriter superstar "Lady Gaga." Also, in Chemistry Block 1, Manar has the Boland nickname "Manar Manar" which is a play off of the song Mah ná Mahná made popular by the Muppets on the children's show *Sesame Street.* Pete has developed a great rapport with students, parents and peers. He's a great motivational speaker, very entertaining, and often tapped to host Beacon's end-of-the-year "Capstone Night[4]." Pete has now put everyone in great spirits before the NECAP testing. It will begin in less than two hours.

8:30 a.m.–9:25 a.m. PRECAP. Consistent with the two previous days of testing, students are participating in one of the three PRECAP activities available: "Walk on Waters", "Fly on the Wall", or, "Muzzle-free Mowry." And, again, the schedule is strictly adhered to: from 9:25–9:45 a.m., it's Brain Food and Bagels with Butler; at 9:45 a.m., students report to assigned testing rooms; between 9:45 and 10:00 a.m. attendance is taken, instructions read, and test materials passed out.

10:00 a.m. Again, work begins. The estimated time for completion for this session is 60 minutes. Everything goes well and by 11:30 a.m., testing is complete. A few students have, however, been absent and will be taking a make-up.

Finally, the pressure of these three days is over. Now, Mrs. Waters and the rest of us will have to wait for about four months until the State releases the results. Nevertheless, we are already discussing preparation for next year's tests.

As this book has taken over a year to write, we have since received the results of Mrs. Water's efforts in preparing her students for the NECAP science assessments. They will be discussed later in this chapter. However, my discussion of them also involves post-NECAP research conducted prior to the end of the 79 days I spent in Mrs. Waters' classroom. The research centered on Mrs. Waters and her chemistry students; and, so represents part of this story.

2:45 p.m.–3:45 p.m.: After-School Professional Development Opportunity. Professional Development (PD) for teachers is time set aside for training. Sometimes PD is mandatory and, sometimes voluntary. I have found that some of the most informative professional development opportunities have come from those led by peers. This year, Mrs. Waters has led two workshops on "Understanding by Design" (UbD). Today, James MacDonald (the school's history teacher) is talking about how he has been using online "discussion boards"[6] in his classroom.

However, in many schools, the top administrators often dictate what PD will be. Being "force fed" information that is not presented in an engaging manner, is generally an unattractive option. Unfortunately, however, it is all too often the case. In fact, I'm sure that my attitude is not unlike those of the many students who get left behind in education because they are being force-fed information that is of little relevance to their futures.

In his 2007 *New York Times* Best Seller, *Words that Work*, author and renown American political consultant and pollster Dr. Frank Luntz advises, "It's not what you say, it's what people hear" (p. xiii). Luntz states that a person on the receiving end of words understands "through the prism of his or her own emotions, preconceptions, prejudices, and preexisting beliefs" (p. xiii). Here is an example that supports Luntz's advice and explains my position on mandatory PD.

It was the beginning of a new school year in 2004, and I was sitting in the windowless library of a fairly large suburban high school along with 196 fellow high school teachers. It was Professional Development (PD) day and the school's administration had been kind enough to schedule a three-hour lecture by an author of a children's book. I suspected the author had been flown in from a distant land at a considerable cost. The title of the book and the rumored topic of the lecture is *What We can Learn from Stuffed Animals* (actually, I don't remember it exactly, but these are the words I heard). Extra chairs have been brought in for the listeners and tables positioned in such a way as to provide as much obstruction as possible should any of us be thinking of escaping! The only discernible means of egress seemed to be crawling on the floor between the legs of the principal who has strategically positioned himself at the main entrance to the library.

Having arrived late, I find myself sitting just one row shy of the back wall. Directly behind me is the entire physical education department. They've already started with catcalls and we're still waiting for the guest speaker to arrive. Finally, the speaker approaches the podium and introduces herself in a very energetic voice:

"Thank you for inviting me here—this is my first time speaking to high school teachers" followed by, "Could you all stand so we can do some warm-up calisthenics."

Considering Luntz, these are not exactly "words that work" for me and here's what I hear and think:

So, let me get this straight, I'm being professionally "developed" by an elementary school teacher who has never spoken to a high school audience, in a room I don't want to be in, with some people I don't want to be with, listening to a topic I'm not interested in; and, I personally find the word *calisthenics* repulsive. Also, the catcallers behind me are growing ever louder. This is hardly good professional development and certainly not conducive to listener involvement.

In contrast to this, there is a PD model that seems to be working quite well at Beacon. Teachers who are knowledgeable and passionate in a particular area take on the role of a presenter and conduct workshops for any of us who wish to attend. For example, in the 2010-2011 school year, five one-hour afterschool workshops were held; two on Understanding By Design (UbD), one on differentiated instruction in the classroom, one on the use of NWEA (Northwest Evaluation Association) data to drive instruction, and one on creating and using Internet discussion boards. I attended all five and was a presenter at one.

MACDONALD'S WORKSHOP ON "USING DISCUSSION BOARDS

James MacDonald (history teacher) has pioneered the use of online "discussion boards" at Beacon. Today, he is leading a workshop on how he has used this in his classroom. He has prepared a nice PowerPoint presentation along with some quality handouts on the topic.

I am immediately impressed by this educational medium and see opportunities to use it both in my classroom and for educational research. I recall the note I had made in my journal to create a post-NECAP survey to investigate student attitudes toward the change of test venue. Students can comment directly on their NECAP science testing experience as to preparation, test environment, and PRECAP activities. In what follows, I discuss the research method and results of exploring students' feelings toward their NECAP preparation with Mrs. Waters and the PRECAP activities during the May 2011 testing.

EXPLORING STUDENTS' ATTITUDES TOWARD NECAP

All eleventh-grade students in public high schools in Rhode Island are administered the NECAP science assessments during May of each school year. The vast majority of students take the "paper and pencil" version in three different sessions. In preparation for the test, all eleventh-graders at Beacon Charter High School receive instruction and practice in their chemistry class. Chemistry is the science requirement for students at this grade level.

In this semester, Spring 2011, Mrs. Jessica Waters is the teacher of record for all eleventh-grade chemistry students. She has three classes (i.e., Block 1, Block 2, and Block 3). During the semester, she has provided several different types of NECAP practice tasks including: POD (Problem of the Day); Study Island with "mistakes documents;" practice inquiry tasks; and, "flexible" homework and projects differentiated on the basis of a student's NWEA weaknesses in various areas of science. Further, students practiced relevant vocabulary in their Senior Prep class.

In Spring 2011, I conducted a study in Mrs. Waters' Chemistry Block 1 class. It is discussed in chapter 7, "A Taxonomy of Time." I there indicate that I determined that there were only 64 days of instruction, not the usual 90 for the semester-long course. This meant that there was a significant reduction in time for teaching the chemistry content of the NECAP as well for practicing for it. There was therefore a major concern as to whether students would be adequately prepared.

As noted, Beacon had already established a model for NECAP test administration on the basis of the NECAP Math, Reading and Writing fall assessments. This included isolating the fourth floor for testing (only test takers were allowed there); a room assignment method based on NWEA scores (the highest achievers were placed in one smaller room—lower achievers in a larger one; food and drinks were made available and there was a pre-testing time called PRECAP which allowed for relaxing activities (e.g., walking, games, a quiet room).

Unfortunately, a controversy arose on the day before NECAP testing. A decision was made, contrary to Principal Dr. Pilkington's wishes, to switch the location from the fourth floor to the third. This was done because one of the fourth floor testing rooms was the Film Room and students who were working on their Capstone films needed it at this time. Dr. Pilkington argued that we had established a testing model that worked. The students to be tested on science in May, 2011 on the third floor, are the same ones who had previously taken the Fall 2010 math, reading and writing assessments on the fourth floor. A change in testing venue might negatively affect their engagement and ultimately impact upon their performance.

Given these facts, I decided to explore student reflections of their NECAP experiences, including their NECAP preparation in chemistry; their recommendations for improvements/changes, if any; their attitudes toward the testing environment; and, their attitudes toward PRECAP. I set out four questions on an online *discussion board* [6]. Mrs. Waters presented the assignment, so none of the students were aware of my involvement in it. In the days immediately following the NECAP test administration, she asked all of the 43 eleventh-grade test-takers to respond to the following four questions she had posted on Beacon's Digication site:

Q1: Do you feel that the NECAP preparation in your chemistry class was adequate?
Q2: Do you have any suggestions for improving NECAP preparation in the future?
Q3: Do you have any comments about the testing environment (e.g., room, lighting, noise)?
Q4: What did you think about the PRECAP activities / refreshments?

I retrieved and compiled all of the students' responses, then constructed a table which reflected those across Mrs. Waters' three chemistry classes. There were 132 postings in total. Also included in the tables were the individual student scoring results that had been made available to Beacon on September 19, 2011.

THE DAY OF RECKONING

Tuesday, September 19, 2011: Admittedly, I did not sleep very well last night. I wouldn't be surprised if Mrs. Waters had a similar experience. After all, today is the "Day of Reckoning." Prior to their pending public release (in about one week), RIDE (Rhode Island Department of Education) is allowing school districts to review their district science scores online at 7:30 a.m. this morning. Here is how all of this plays out:

6:20 a.m. I sign into school.

7:10 a.m. Academic Dean, Mike Skeldon, has his computer ready.

7:15 a.m. I join Mike in his office with my breakfast bran muffin.

7:17 a.m. He complains about muffin crumbs on his desk.

7:25 a.m. Mrs. Waters arrives and joins us in Mike's office. She's noticeably nervous. I indiscriminately push crumbs onto the floor and sweep them under Mike's desk with my shoe.

7:30 a.m. Mike types in the special password for Beacon issued by RIDE to individual schools. Good News!

After two years of stagnant performance, we have doubled our proficiency score—up from 14% in 2010 to 28% in 2011. Scanning the results further reveals that we have beaten the state average in all of the four assessed areas of science: *Physical Science*, *Earth & Space Science*, *Life Science*, and *Scientific Inquiry*.

Of the 43 test takers, only 9% are deemed "Substantially Below Proficient," 63% are "Partially Proficient," 28% are "Proficient," and none are "Proficient with Distinction." The current State requirement for graduation is a minimum achievement of Level 2 (Partially Proficient) on the math and reading assessments—there is no such State requirement for science at this time. However, if the same benchmark were in place, 91% of Beacon's eleventh-graders have met or exceeded the requirement of "Partially Proficient."

Of course, the three of us agree there is plenty of room for improvement, but these statistically significant positive results certainly suggest that Mrs. Waters' effort and methods, and our model for NECAP administration, seems to be working. Now that these results had been released, I could explore student feelings and attitudes toward the process in so far as these were linked to their NECAP performance. The following discussion is based on analyses from two different sources of student data ("self-reported") as it relates to their feelings and attitudes toward NECAP. These sources were the online discussion board and the NECAP Student Answer Booklet questionnaire.

Results from the Online Discussion Board

Of the 43 test-takers, 33 responded to the four questions posed on the online discussion board. This equated to a 77% response rate. There were a total of 132 postings. I was particularly interested in those student responses that I deemed worthy of consideration because they could affect future decisions on NECAP preparation and test administration. Tables 6.1 through 6.4 provide a brief summary of results along with key student responses. I have also included their NECAP four-point rubric score together with the corresponding "proficiency level" they had achieved. Both are placed within parentheses (Note: Level 4 = Proficient with Distinction; Level 3 = Proficient; Level 2 = Partially Proficient; and Level 1 = Substantially Below Proficient).

Some Conclusions from the Online Discussion Board

Overall, students feedback on their NECAP experience was positive:

- For all practical purposes, there was unanimous agreement that they felt Mrs. Waters had adequately prepared them for the science exam.

Table 6.1. Online Discussion Board Results for Question One

Q1. Do you feel that the NECAP preparation in your chemistry class was adequate?
Results
97 percent (32/33) of the respondents stated they had been "adequately" or "more than adequately" prepared by Mrs. Waters for the NECAP science assessments (see response "3" below).
Key Student Responses
1. "Yes, the chemistry portion of the test was easy because of the PODs we've been doing, but the biology and astronomy portion was more difficult (but still doable)" (achieved Level 3—Proficient). 2. "Yes, it was a lot better than Study Island. A lot less stressful as well" (achieved Level 2—Partially Proficient). 3. "Not really because I was absent the whole week that all the students prepared but when I came back, I was given the opportunity to study for one day instead of taking the NECAP. So I just had a make-up day" (achieved Level 3—Proficient). 4. "Yes and No. We spent a lot of time beforehand preparing but I felt it was too rushed and took away from class that could have been spent helping student learning. I think students should have to do NECAP prep work from freshman year on (All NECAPs, not just science) so it's not all rushed and students can be more prepared and not lose so much class time" (achieved Level 2—Partially Proficient).

- Despite the fact that these test-takers had been located on the fourth floor for the Fall 2010 NECAP math, reading, and writing assessments, and then moved to the third floor for the Spring 2011 NECAP science assessments, only one student mentioned in passing the change in venue. ("The room was a little messy. We should have done the same exact thing for the reading and math" (see table 6.3)). I'm assuming that what was meant here was ". . . the same exact thing *as* reading and math" not that we should have had a messy room when we took the reading and math exams last fall. Further, as concerns Question #3 (on testing environment), all of the negative comments that were made might have applied to either the third or fourth floors. Therefore, considering the results of the NECAP science assessments, i.e., the doubling of the percent of students identified as "proficient" over the previous year, any adverse effects caused by the change in venue are probably statistically negligible. However, considering that some students fell short of proficiency by one or two points, attention should be paid to the references made in their responses to distractions.
- The vast majority of students (over 85%) reported that they had had a positive experience with PRECAP, and several specifically mentioned that it

Table 6.2. Online Discussion Board Results for Question Two

Q2. Do you have any suggestions for improving NECAP preparation in the future?
Results
55 percent (18/33) of the respondents did not feel the need to offer suggestions for improving NECAP preparation or the suggestions/comments offered were deemed not worthy for further discussion. Fifteen student responses were highlighted for further discussion. 73 percent (11/15) of these students received a "2" (Partially Proficient) and 27 percent (4/15) received a "3" (Proficient) as noted: Key Student Responses 1. "I think we should have focused on the vocabulary words we did in our senior class. I felt like we were rushed through them and had no time to study and understand them" (achieved Level 2—Partially Proficient). 2. "I do not believe that there could have been more preparation. 'Problems of the Day' and practice inquiry tests definitely benefitted me in the long run" (achieved Level 2—Partially Proficient). 3. "Spend more time prepping for inquiry tasks" (achieved level 2—Partially Proficient). 4. "Overall, the NECAP preparation was really good. The one problem I saw was that we didn't focus a lot on vocabulary, we just rushed through it". (achieved Level 2—Partially Proficient). 5. "Just add more focus on biology and astronomy to the chemistry prep" (achieved Level 3—Proficient). 6. "More on genetics and earth and space science. For chemistry, maybe more faster so that more lessons can be all covered" (achieved Level 3—Proficient). 7. "More Study Island topics, Study Island is very helpful" (achieved Level 3—Proficient). 8. "Don't go so fast also don't pile so much on us last minute" (achieved Level 2—Partially Proficient). 9. "One example the future chemistry class can do for preparation for the science NECAP is do NECAP packets for at least three weeks straight so it can get drilled in their mind" (achieved Level 2—Partially Proficient). 10. "Some future preparations can be having science workshops on the weekends for some students who would like to attend" (achieved 2—Partially Proficient).
Q2. Do you have any suggestions for improving NECAP preparation in the future?
Key Student Responses
11. "I feel that the sophomores should start preparation at the start of their junior year; so that there will be more of an understanding and time for those who don't understand to improve" (achieved Level 2—Partially Proficient). 12. "I think the only thing I think should happen is to start prepping earlier than we did" (achieved Level 2—Partially Proficient). 13. "I kind of just explained it up there in Question 1. But, let's repeat. Start NECAP prep work from Day 1 to the end of the school year so we won't lose a lot of class time and students will be more prepared. Do a lot of inquiries" (achieved Level 2—Partially Proficient). 14. "I think personally, next time, if we worked on understanding graphs better because some of the graphs found on NECAP were hard to understand" (achieved Level 2—Partially Proficient). 15. "I think they should start prepping for NECAP in the beginning of the year along with Math and English" (achieved Level 3—Proficient).

did help them relax before the test. The following three response typify the overall attitudes of the test takers toward PRECAP:

> I love the fact that the school does the walks and gives students a chance to relax before they take the test. The quiet room is a great way for students to go and read or listen to music and just zone out. The hangout room was another great way for students to relax, talk to friends, and just relax. Most schools don't do that and I love that this school does. Also, having the different options definitely helps. That way, students can relax in a way that is best for them (achieved Level 2—Partially Proficient).
>
> The PRECAP activities were good and I think that it was very generous of the teachers to put as much time as they did with the preparation of the testing as well as the act of making us feel comfortable (achieved Level 3—Proficient).
>
> That is my favorite part of NECAP. It gets me relaxed and ready for the actual NECAP (achieved Level 2—Partially Proficient).

Some Recommendations Based On the Online Discussion Board Analysis

The following recommendations were forwarded to Mrs. Waters and the Beacon administration for consideration as regards future testing:

1. A few students noted that they perceived that their teachers were stressed by this testing and they claimed that this overflowed to the students. We need to make sure that teachers do not exhibit this behavior either verbally or with body language.
2. A student noted that she felt stressed by teachers hovering over her. We need to develop an addition to the instructional protocol, one that is read to the students at the beginning of the test session and explains that teachers will periodically walk around the room to get an idea of where students are relative to the allotted time. Students need to be clear that teachers are not doing this for the purpose of checking answers. Or, maybe we could develop a system of index cards that students can display on their desk when they have completed half the questions.
3. One student noted that she had lost motivation to work to the best of her ability when other students finished and got up and left. We need to decide how we are going to handle students who finish early as well as how we accommodate the lunch period as these often coincide.

Table 6.3. Online Discussion Board Results for Question Three

Q3. Do you have any comments about the testing environment (e.g., room, lighting, noise)?
Results
55 percent (18/33) of the respondents were satisfied with the environment and either did not feel the need to make any additional comments or made no comment at all. Fifteen student responses were highlighted for further discussion. 13 percent (2/15) of these students received a "1" (Substantially Below Proficient): 53 percent (8/15) of these students received a "2" (Partially Proficient); and, 33 percent (5/15) received a "3" (Proficient) as noted:
Key Student Responses
1. "It could help with maybe some background noise, music or such, to help keep some awake and also because it helps some students, such as me to focus. Heating could also be fixed as I found it a tad chilly" (achieved Level 3—Proficient). 2. "In Mr. Boland's room, after 11:10 am, the students who just got testing in the computer room were really loud and you could hear everything in which made me lose focus every now and then" (achieved Level 3—Proficient). 3. "I feel as if the hovering that the teachers were doing put a lot of unnecessary stress on me. I don't like that feeling over someone standing over me and evaluating me as I work. I think that next year the teachers should just sit around the room and come if we ask for their assistance" (achieved Level 3—Proficient). 4. "No, I enjoyed being in a room with a small amount of kids testing though. At times it seemed as if different teachers would go over the same topics, which grew to be an annoyance, but having teachers like Ms. Fox and Mrs. Waters going over the work only would allow the kids to learn without over doing it" (achieved a Level 2—Partially Proficient). 5. "Yes, I think the testing environment could use a bit more room or less people in the room because I found it crowded. Or move the tables a different way next time" (achieved Level 1—Substantially Below Proficient). 6. "The room was a little messy. We should have done the same exact thing for the reading and math" (achieved Level 2—Partially Proficient). 7. "I did not have any problems with the testing environment. Choosing a large room made it easier to spread out and take the test at your own pace instead of instinctively competing with the person sitting next to or across from you. I liked taking it in the Theatre room" (achieved Level 2—Partially Proficient).
Q3. Do you have any comments about the testing environment (e.g., room, lighting, noise)?
Key Student Responses
8. "I liked taking my test in the theatre room. It was big but it was quiet and easy to focus" (achieved Level 2—Partially Proficient). 9. "I most definitely think that being put into a smaller room most definitely helped to keep my concentration and focus on my test. The only thing that distracted me was when classes were changing" (achieved Level 2—Partially Proficient).

10. "My only complaint was before the actual testing. Some people need to mind their own business and not be concerned with what other people are doing. Some of the girls were acting like bitches. It was probably because of the testing. Excuse my language. During the actual testing though, everything was fine. I was comfortable which I guess means that most in this kind of testing for the best test performance" (achieved Level 2—Partially Proficient).
11. "I think the environment was perfect except it was a little stuffy" (achieved Level 3—Proficient).
12. "The testing environment was fine. I just felt that every time it came 11:10 and people started to leave, that my interest in finishing the test to my best ability just went downhill" (achieved Level 2—Partially Proficient).
13. "I felt that the testing environment was good, just a little crowded" (achieved Level 2—Substantially Below Proficient).
14. "Yes, I do thanks for asking. Some ANONYMOUS teacher was eating chips and opening cans of ARNOLD PALMER. I was distracted and now the school will be shut down for insufficient scores" (achieved Level 3—Proficient).
15. "The testing environment for the NECAP test was OK. The room was very big and the noise would travel around the room. If someone was talking from one side of the room you would hear it from the other side. Although that was one major downfall, everything else was fine" (achieved Level 2—Partially Proficient).

4. Two students suggested increasing opportunities to practice relevant vocabulary words in the Senior Prep classes prior to taking the NECAP.
5. Several students recommended starting NECAP prep in still earlier grades; one suggested the addition of weekend workshops.
6. There were requests for more practice on inquiry tasks, genetics, Earth & Space, astronomy, biology, and use of Study Island.

STUDENT RESPONSES TO THE QUESTIONNAIRE CONCERNING THE TEST ANSWER BOOKLET

At the end of the three NECAP assessment sessions, test takers are asked to complete a questionnaire in which they answer 19 multiple-choice type questions. They cover, for example, their feelings towards the difficulty of the test, use of calculators, and science courses that they are currently taking. Students are asked to "bubble-in" their responses on the back of the Answer Booklet they used for the science assessment. So, their responses are directly linked to their performance. The following are the results of this questionnaire (see table 6.5). Also included is a brief discussion of findings for the 43 Beacon eleventh-graders who took the NECAP in the Spring 2011. A

Table 6.4. Online Discussion Board Results for Question Four

Q4. What did you think about the PRECAP activities / refreshments?
Results
Overall, about 85 percent (28/33) of the respondents enjoyed PRECAP.
Key Student Responses
1. "I think the walks were pretty fun and the breakfasts were okay. They were a bit better last October when we took the math and English NECAPs" [achieved Level 2—Partially Proficient]. 2. "Because I was going to Dunkin' Donuts, I was enjoying it. I don't know about the food that was being served at the school because I didn't try any of it other than fruit. However, I felt like the time that was given to us before NECAP was really long. I was getting bored" (achieved Level 2—Partially Proficient). 3. "I did not like that we had to stay in the cafeteria to talk. It felt crowded and dirty, but Ms. Wall's room was fine" (achieved Level 3—Proficient). 4. "I enjoyed the PRECAP activities and refreshments but some of the teacher's stress level was making me nervous and stressed when taking the NECAPs" (achieved Level 2—Partially Proficient).

comparison between the results for Beacon students and those for the State as a whole is as well.

Some Conclusions from the Student Questionnaire on the Answer Booklet

The following is a brief summary of the key findings from my analysis of the student responses to the 19 questions on the back of their answer booklet. As noted, they answered these at the immediate conclusion of the last day of testing. Statewide data was provided by RIDE (Rhode Island Department of Education):

- Beacon students consider their class work to be as rigorous as the NECAP test.
- The average non-Beacon student (i.e., according to State statistics) is not being sufficiently challenged in their science classroom.
- Beacon students feel more prepared for NECAP than non-Beacon students.
- 100% of Beacon students reported they tried at least as hard on the NECAP test as they do on their science class work; this stands in contrast to 78% of non-Beacon students (22% of the state students reported they did not try as hard on the NECAP test as they did on their science class work).
- There were no significant differences between Beacon students and non-Beacon students as to how well they understood the test directions. The

Table 6.5. NECAP Student Questionnaire Results

	Question / Question Response Choices	*Beacon n = 43*	*State N = 14,669*
1.	How difficult was this science test?		
	a. It was harder than my regular science schoolwork.	24	43
	b. It was about the same as my regular science schoolwork.	61	34
	c. It was easier than my regular science schoolwork.	15	15
2.	How hard did you try on this science test?		
	a. I tried harder on this test than I do on my regular science schoolwork.	59	18
	b. I tried about the same as I do on my regular science schoolwork.	41	54
	c. I did not try as hard on this test as I do on my regular science schoolwork.	0	20
3.	How well did you understand the directions your teacher gave you when you were taking the science test?		
	a. I understood the directions very well.	93	79
	b. I needed some help understanding the directions.	7	8
	c. I needed a lot of help understanding the directions.	0	4
4.	Did you have enough time to answer the questions on sessions 1 and 2 of this science test?		
	a. I had enough time to answer the questions and check my work.	85	77
	b. I had enough time to answer the questions, but I did not have time to check my work.	12	9
	c. I felt rushed, but I was able to finish answering the questions.	2	3
	d. I did not have enough time to finish answering the questions.	0	2
5.	How often did you use the reference sheet?		
	a. I used it on all of the questions that suggested it even for some questions that didn't.	22	10
	b. I used it on most of the questions that suggested it.	24	20
	c. I used it on some of the questions that suggested it.	37	23
	d. I didn't use it on very many questions.	17	36
6.	How much did you use a calculator on the test?		
	a. I used it on most of the questions.	2	2
	b. I used it on some of the questions.	10	8
	c. I didn't use it on very many questions.	44	21
	d. I didn't use a calculator but wanted to.	0	8
	e. I didn't need a calculator.	49	53
7.	Did you have enough time to answer the questions on the inquiry task?		
	a. I had enough time to answer the questions and check my work.	83	71

Table 6.5. *(Continued)*

	Question / Question Response Choices	*Beacon n = 43*	*State N = 14,669*
	b. I had enough time to answer the questions, but I did not have time to check my work.	15	13
	c. I felt rushed, but I was able to finish answering the questions.	2	5
	d. I did not have enough time to finish answering the questions.	0	2
8.	How often do you do inquiry tasks like the one that you did on this science test?		
	a. a few times a week	49	12
	b. a few times a month	41	32
	c. a few times a year	5	21
	d. never or almost never have inquiry tasks like this	5	26
9.	How often do you work with other students on science experiments or inquiry tasks?		
	a. always	27	34
	b. sometimes	66	44
	c. never	7	8
	d. My teacher usually does the science experiment or inquiry task and we watch.	0	2
	e. We do not do science experiments or inquiry tasks in my class.	0	4
10.	Which of the following best describes the science class you took in grade 9?		
	a. Physical Science	74	53
	b. Biological Science	10	13
	c. Earth Space Science	5	14
	d. General or Integrated Science	10	9
	e. I did not take a science class in grade 9.	0	1
11.	Which of the following best describes the science class you took in grade 10?		
	a. Biological Science	85	66
	b. Earth Space Science	2	7
	c. Chemistry	7	12
	d. Physics	2	3
	e. I did not take a science class in grade 10.	2	3
12.	Which of the following best describes the science class you are taking in grade 11?		
	a. Biological Science	0	11
	b. Earth Space Science	0	6
	c. Chemistry	95	47
	d. Physics	2	11
	e. I am not taking a science class this year.	2	16

	Question / Question Response Choices	Beacon n = 43	State N = 14,669
13.	Which of the following best describes the science class you plan to take in grade 12?		
	a. Biological Science	47	14
	b. Earth Space Science	23	7
	c. Chemistry	0	13
	d. Physics	19	29
	e. I do not plan to take a science class in grade 12.	13	27
14.	How often do you have science class this year?		
	a. I have science class every day.	68	38
	b. I have science class on a rotating schedule (3–4 times per week).	3	16
	c. I have science class less than three times per week.	0	3
	d. I have science on a block schedule.	30	19
	e. I am not taking science this year.	0	16
15.	How often do you have science homework this year?		
	a. every day	17	20
	b. a few times a week	83	39
	c. a few times a month	0	9
	d. I usually don't have homework in science.	0	8
	e. I am not taking science this year.	0	16
16.	What was your science grade on your most recent report card?		
	a. A or equivalent	40	25
	b. B or equivalent	33	38
	c. C or equivalent	18	19
	d. lower than C or equivalent	10	8
17.	In your most recent science class, how often do you create tables, diagrams, charts, or graphs to represent data?		
	a. frequently	59	33
	b. sometimes	37	37
	c. rarely	5	14
	d. never	0	5
18.	In your most recent science class, how often do you present the results of your investigations?		
	a. frequently	63	29
	b. sometimes	34	32
	c. rarely	2	19
	d. never	0	10
19.	Do you use a science journal or science notebook to write about your thoughts and experiences in science class?		
	a. yes	71	22
	b. no	29	65

majority of the members of both groups reported they understood the directions very well.

- There were no significant differences between Beacon students and non-Beacon students as to their feelings about the time allotted to answer the questions in the first and second sessions; similarly, for their feelings about having enough time to complete the Inquiry Task in the third session. The majority of members of both groups reported that they had had enough time to answer their questions and check their work.
- There were no significant differences between Beacon students and non-Beacon students as to use of calculators. Approximately 50% in each group reported they did not need a calculator.
- Beacon students reported having done more practice work on the Inquiry Tasks than non-Beacon students. Recall that Mrs. Waters had specifically built practice inquiry tasks into her curriculum during the semester.
- The majority of both Beacon and non-Beacon students reported that they had followed the same science course sequence in high school, taking physical science in ninth grade, biological science in tenth, and chemistry in eleventh.
- The number of Beacon students who reported that they had science classes every day was significantly greater than that of non-Beacon students.
- Beacon students reported having more science homework than non-Beacon students.
- Beacon students reported that they received more "A's" in their science class than non-Beacon students.
- Beacon students reported that they had more opportunities to present their investigations and write about their learning experiences in science classes.

BEACON VS. SENDING DISTRICTS ON NECAP SCIENCE ASSESSMENTS

RIDE expects charter schools to outperform those of their contributing districts (i.e., sending school districts that Beacon students would have attended had they not enrolled at Beacon) and also outperform schools that serve similar student populations. Performance is measured primarily by NECAP tests. The overall target achievement score that RIDE sets for Beacon is a sum of separate weighted calculations for each sending school district. It is based on the product of the percent of Beacon test takers from a particular sending district and the NECAP performance of that particular district (more specifically the high school). For example, if there are 50 test takers at Beacon for a particular testing year, and five students were from Providence, then 10% (5/50) of Beacon's students must achieve or surpass the NECAP performance level of the Providence school that the students would otherwise have attended. This weighted calculation is performed for all of Beacon's contributing school districts. The sum of all of the weighted calculations is then set as Beacon's target score. The use of NECAP results (RIDE 2011a), as the basis for charter school performance, is presented in RIDE's Charter School Performance Review Handbook (Butler 2011a). Further, all schools typically compare their NECAP performance to the State's average so as to gauge the school's relative performance.

The following research question was developed as a way of quantifying Beacon's longitudinal NECAP performance as compared to that of its contributing districts and other Rhode Island high schools: "How does Beacon's performance on the NECAP science assessment compare with that of its contributing school districts and with Rhode Island's high school average for the testing years 2008 through 2012?

Addressing the Research Question

As shown in table 6.6 and associated figure 6.1, students at Beacon Charter High School have outpaced both its contributing school districts and those of the State over the last four years where proficiency percentages are concerned. Since the 2008–2009 school year, interestingly, the year Mrs. Waters was hired, Beacon with its 39.9% gain has outpaced both its contributing districts (a 12.3% gain) and the State (a 15.6% gain) in terms of percent growth in NECAP science proficiency performance. The school now has the second highest four-year growth on the NECAP science assessments of all of the 53

Table 6.6. Comparisons of NECAP Science Percent Proficient (and above) for Testing Years 2008-2012

Test Date	*Beacon*	*Sending District Average*	*State (H.S.)*
May, 2008	4.8 ± 4.6	10.4	16.4
May, 2009	13.7 ± 4.8	17.2	19.4
May, 2010	14.3 ± 5.4	11.0	20.5
May, 2011	27.9 ± 6.8	21.8	25.7
May, 2012	44.6 ± 6.6	22.7	32.0
4-Year Growth:	+ 39.9	+ 12.3	+ 15.6

Note: Beacon Charter High School has the 2nd highest 4-year growth on the NECAP Science assessments for years 2008–2012 out of 53 Rhode Island high schools. Only Scituate High School at 42.8% was greater.

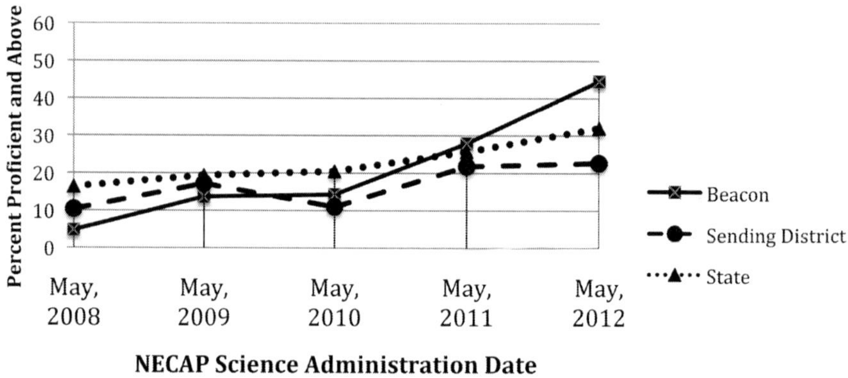

Figure 6.1. Beacon's NECAP science performance (2008–2012).

Rhode Island public high schools that took the NECAP. Only Scituate High School (a 42.8% gain) is greater (RIDE 2011a).

Major Reasons for Beacon's Success on the State Science Assessments

Based on my 79 days of participant observation research in Mrs. Waters' Chemistry Block 1 class, a period coincident with the preparation for and administration of the State NECAP Science assessments, I would propose that the following were the major factors in Beacon's success:

- Targeted and intensified instruction by Mrs. Waters, the eleventh grade chemistry teacher, focusing on NECAP based standards (i.e., the science GSEs)
- Her POD (Problem of the Day) based on RIDE-released items
- Mrs. Waters' use of the Study Island web-based software tool
- Her practice Inquiry Tasks built into the science curriculum
- Increased familiarity on her part with the NECAP science assessments
- Her use of NWEA science scores to differentiate instruction on homework and project assignments
- Her regularly soliciting student feedback through surveys and questionnaires; online discussion boards; reflections, etc. to assess instruction and solicit comments on the testing environment
- Beacon's establishment of a culture where all students try their hardest on the assessments
- Its focus on providing a comfortable testing environment for students
- Its creation of a pre-testing environment (i.e., PRECAP) intended to defuse student anxiety

CLOSING COMMENTS ON MRS. WATERS AND THE STATE SCIENCE ASSESSMENTS

In her January 8, 2013 *The Washington Post* article entitled "Gates Foundation study: We've figured out what makes a good teacher," Lindsey Layton reported the findings of a $45 million research project by the Bill and Melinda Gates Foundation that involved 3000 teachers and their students. It determined that the most reliable way to evaluate the effectiveness of teachers was by using a three-pronged approach—from student test scores, classroom visitation by multiple observers, and teacher evaluations by the students themselves (Layton 2013). So, I am very proud to be able to state that Mrs. Waters' efforts were impressive, as were the results of the NECAP Science assessments. My 79 consecutive days of classroom observation and participation in her chemistry class, together with our collaborative solicitation and use of student evaluations support this same conclusion. And, we did not have to draw on a budget of $45 million to do this!

NOTES

1. John D. Rockefeller (1839–1937). American Industrialist and philanthropist. This quote was found in *Success Library, Volume 9*, (1901, p. 1430) edited by Orison Swett Marden and George Raywood Devitt (New York: The Success Company Publishers).

2. The Common Core State Standards (CCSS) initiative is a state-led effort coordinated by the National Governors Association Center for Best Practices (NGA Center) and the Council of Chief State School Officers (CCSSO). These standards were developed in collaboration with teachers, school administrators and experts in an effort to provide a clear and consistent framework to prepare our children for college and the workforce. They are based on best practices in national and international education, and also on research and input from numerous sources. On June 2, 2010, the final version of the CCSS for English Language Arts (ELA) and mathematics was released to the public. A majority of states (45 as of this writing) have agreed to adopt the CCSS.

3. ASCD (formerly the Association for Supervision and Curriculum Development), a nonprofit, nonpartisan membership association, was founded in 1943 as an educational leadership organization dedicated to advancing best practices and policies for the success of each learner. ASCD located in over 145 countries, has 150,000 members comprised of professional educators from all levels and subject areas—superintendents, supervisors, principals, teachers, professors of education, and school board members.

4. Beacon students must complete a Capstone Project in their senior year as a graduation requirement. Every senior takes a film class and must produce a film, tak-

ing it from the storybook stage all the way through the filming and the editing process. The students present their work before a panel of judges. The process culminates in Beacon's Capstone Night, held annually at the end of the school year. This major event is modeled on Hollywood's *Academy Awards* (the Oscars) for excellence in cinematic achievement.

5. *Muguet* is the French name for the poisonous woodland flowering plant more commonly known *Lily of the Valley*.

6. An online *discussion board* is a website on which people can hold conversations in the form of posted messages. These can be viewed in real time or at some future point. It is a type of internet forum and can be a useful tool for teachers who are trying to promote peer collaboration and to qualitatively assess students' learning.

Chapter Seven

A Taxonomy of Classroom Time

"Don't count every hour in the day, make every hour in the day count."

—Alfred Binet

Taking on the role of a student in Mrs. Waters' class in the Spring of 2011 was not much of a stretch for me. I have spent most of my life as a student from elementary school to the PhD in education—26 years all told. Many of those years were spent squeezing my endomorphic body between a desk and the back of its hard polypropylene seat. So much squeezing that summertime at the beach is always problematic. Some of the opposite gender worry about their "bikini line," but I worry about my "desk line!"

During those many hours of sitting as a student, my mind has run the full spectrum of activity and non-activity, from being highly engaged and fully awake to stage-five sleep (i.e., REM). At some points along that continuum, when I have become disengaged from classroom work, mischievous thoughts have been apt to appear. In fact, the research supports the conclusion that behavior problems are most likely to occur when students are not engaged in some activity, i.e., at either the very beginning or the very end of class. One class in my doctorate program met once a week. The professor was notorious for not taking a break midway through, as doing so was customary in other classes. On one particular evening, at the start of class, the professor announced to our cohort of twelve students that he wanted to take a break at the midpoint. We all looked at each other in shock. This was unprecedented! Just before the halfway point, when I was admittedly distracted from the lecture and sensed that the professor was about to signal the break, a mischievous thought entered my mind. I raised my hand. "Yes, John?" he asked. "Professor, before we take our break, I have a question." From both sides of the room there were daggers gleaming from the eyeballs of my peers. "Professor, when does a theory become a theory?" I asked. I knew full well that this was one of those philosophical questions that would not have a short answer. "John, I'm glad you asked that question." Subsequently, he took the next hour and a half to address my question and we never did get to take the promised break. I didn't earn many points with my peers that evening but the event has since provided fodder for discussion among us. I tell this story for two reasons: first, I misbehaved as a student by being disengaged from the flow of classroom events; and secondly, I could have easily replaced my "theory" question with another one, e.g., "What is time?"

This question has been contemplated by some of the most brilliant philosophers, scientists, and mathematicians for over 2500 years including Socrates, Zeno, Plato, Aristotle, St. Augustine, St. Thomas Aquinas, Bacon, Descartes, Barrow, Spinoza, Newton, Leibnitz, Kant, Hegel, Bradley, Poincare, Whitehead, Minkowski, McTaggart, Reichenbach, Wittgenstein, Milne, Dirac, Gödel, and Dummett to name few. However, what has developed over the ages by way of a response to it is an impressively long list of still more questions, opinions, paradoxes/contradictions, hypotheses, conjectures, and theories. Needless to say, it is guaranteed to stimulate debate at either family gatherings or conclaves of metaphysicists. A comprehensive historical discussion of "time" is provided by Bradley Dowden of California State University in his appropriately named article "Time" (Dowden 2010).

Although defining time has proven problematic, the motivation for doing this is not. As Dowden (2010) states "We use our concept of time to place events in sequence one right after the other, to compare how long an event lasts, and tell when an event occurs."

In reading Dowden, I discovered something I had never really thought of before. But the connection to classroom time was immediately obvious to me. There are two different types of time, *physical* and *psychological*. He provides the following definitions: "Physical time is public time, the time that clocks are designed to measure." The 180-day school year 6 ½ hour school day, and an 80-minute-long class are examples of this. "Psychological time (which is called phenomenological time or brain time) is private time." The duration of time that a student is engaged in his/her learning is an example of psychological time.

An important distinction between the two is that physical time can be related to repetitive and consistent intervals at which natural events occur, e.g., Earth's rotation about its axis. For all practical purposes, the time interval created by each rotation is constant. However, with psychological time, human thought processes vary relative to physical time. How many times have you thought or heard someone say "This day is really going by slowly" or, conversely, "This day is really going by fast." Such variation happens in the classroom as well. From a student's perspective, boredom and the lack of understanding are two states of mind that can slow time down, while full engagement in the learning process can speed it up.

Where education is concerned, what is problematic is the lack of attention given to the notion of psychological time in the research literature. If we, as educators, agree that individuals learn at different rates, is it ethical to give timed assessments or to hold teachers and their students to strict curriculum pacing guidelines? Does it matter if a geometry student takes 20 minutes to successfully prove the *Alternating Interior Angle Theorem* when another student can do it in 10 minutes? Or, does it matter if a chemistry student can balance a chemical equation in 15 minutes whereas it takes another one five? How do we measure psychological time and how exactly does it relate to student achievement?

As I get older, and the number of students I have taught increases. Remembering names on the spot is becoming ever more difficult. I recognize faces, recall the interaction with that person, but just can't put the name to the face. Sometimes, it will be minutes, hours, or even a day or two later, before it comes back to me. Eventually, it will pop into my mind even though I was not consciously trying to solve the mystery. Apparently, my subconscious mind continues to work on the problem—analogous to computer security software that is always running in the background. It remains hidden from view and allows the operator to work on something else.

The famous French mathematician and philosopher Jules-Henri Poincaré (1854–1912), who wrote on the subject of intuition and problem solving, believed that fruitful ideas were generated in the conscious state, but that final choices are made unconsciously (i.e., in the background). In the subconscious state, we select amongst a multitude of combinations that will produce positive feelings with regards to such subjective concepts as balance, beauty, clarity, elegance, harmony, proportion, symmetry, and unity. Only when a combination satisfies one's aesthetic feelings will an idea move into consciousness in the form of an inspiration (King 1992), i.e., the "Aha!" moment.

Given this apparent ability of the human mind to subconsciously pursue the solution to a problem beyond some pre-defined "end time" (e.g., the period allotted for a quiz or test), this idea of psychological time remains troublesome. It clouds the picture when it comes to truly assessing a student's problem-solving ability in terms other than mere efficiency of recall.

In other words, does the transmission rate of data in the conscious state vary depending on the background problem solving that is simultaneously occurring? Is there such a metric as "BBS" (brain-bits-per-second)? . . . wow!

SYNCHRONIZING THE SCHEDULE FOR EDUCATION WITH OTHER LIFE RESPONSIBILITIES: A BRIEF HISTORY

Although measuring psychological time may ultimately require the application of electrodes to the brain and/or MR imaging, defining physical time in an educational setting is relatively straightforward—a calendar and a clock suffice.

The school year wasn't always 180 days (September to June) and a long summer break. In fact, during the 1800's, in both rural and urban areas, students spent summers in school. However, the length of the school year and start-and-end dates varied widely between rural and urban areas. Generally, rural communities had much shorter school years. They had a start-and-end timeframe that varied depending on local agricultural work requirements. In contrast, schools in urban areas were open year-round. However, in the last half of the nineteenth century, an interesting phenomenon occurred, namely, the convergence of the lengths of rural and urban school years toward (but only approximating) a standard 180-day school year with a long summer vacation.

Several economic, social and demographic trends triggered and sustained the paths towards convergence on a standard school calendar. Two detailed resources on the subject of school calendar reform are William A. Fischel's *Making the Grade: The Economic Evolution of American School Districts* (2009); and, Kenneth M. Gold's *School's In: The History of Summer Education in American Public Schools* (2002). Although both authors differ in the emphasis they place on the various factors that influenced school calendar reform, these two texts taken together as a whole, provide a rich history and comprehensive list of reasons why the 180-day school year eventually evolved in the twentieth century. The following is a brief summary of some major influences from both the rural and urban perspectives.

In the rural areas throughout the United States in the mid 1800's, the school year generally consisted of two sessions—three or four months in the summer and winter. However, there was a huge disparity in semester length and in the start-and-end dates for these sessions. Duration and start/end dates were linked to local agricultural work requirements—the greater the degree to which the local economy depended on agriculture, the shorter the school year would be. Thus, variation in the start-and-end dates of the school sessions was mostly attributable to local climate.

The summer and winter school sessions made sense given that, for most farming regions in the US, children had to be available to help with the spring planting and fall harvesting. So, what motivated rural schools to replace the summer session with a long vacation and extend the school year? Fischel

(2009) argues that the transition to age-graded education system was the primary impetus for the adoption of a standard school calendar. Somewhere between 1890 and 1910, age-graded education became the national norm. One room schools were gradually replaced with multi-room structures to produce the type of school buildings we have today. With age-grading, students were expected to learn the same lessons at the same time. Therefore they all had to arrive at school at the same time and attend classes on the same days. Further, Fischel states that a long summer vacation facilitated labor mobility. Families and teachers sometimes needed to relocate from one part of the country to another and be ready on the same day as other children were for the start of a new school year.

This explanation is highly plausible when one considers that there was a significant change in the makeup of rural and urban America. It began in the latter part of the late nineteenth century with the shift from the countryside to the cities. During the twentieth century, the urban population in the US doubled from 39.6% to 79.2%. For the same 100-year period, the rural population was reduced by two-thirds from 60.4% to 20.8%. The two trends intersected around 1918 when the urban and rural populations matched (see fig. 7.1).

Fischel also discusses another influence that favored the development of the September–June school year. Colleges and universities had already converged to the standard calendar by the late nineteenth century with June graduation being the norm. Fischel suggests that this model may then have filtered down to the high schools and lower grades.

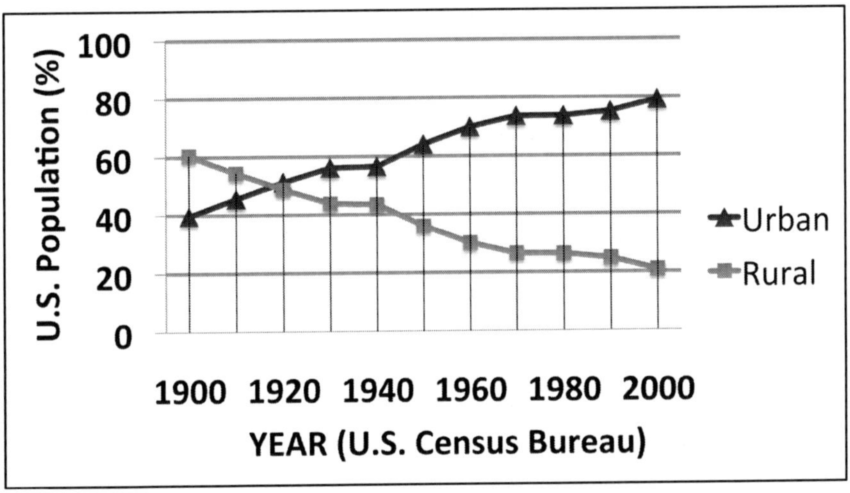

Figure 7.1. Urban and rural population trends (1900–2000).

As concerns the urban areas throughout the US in the mid-1800's, urban schools were open year-round. Their school years averaged nearly 250 days and extending through the summer. As Gold (2002, 71) states: "Summer was universally once an integral part of the urban school year with well-attended schools and substantive curricula." However, in the last half of the nineteenth century, there was a significant trend toward decreasing the length of the school year, reducing it to its current 180-day standard. This occurred in the major cities major cities as shown in table 7.1.

Gold (2002) argues that it was a growing public belief about the frailty of the human mind and body that led to the adoption of the long summer vacation. He cites Horace Mann, who was appointed the first secretary of the Massachusetts Board of Education in 1837, as a strong advocate for the systemization of schooling and the standardization of the school calendar. It would extend the school year in rural areas and reduce the year-round calendar in the urban areas because it included a long summer vacation. Mann believed the long summer vacation was necessary to "avoid the mental and physical fatigue of students and teachers alike," (p. 79).

Today, most states have legislated a minimum of 180 days of school attendance. Some school calendars include a few days beyond this. These extra days are dedicated to teacher training (e.g., professional days); and student interventions (e.g., summer school remediation programs). In fact, in many US school districts, including those in Rhode Island, teacher and student summer vacations are being encroached upon ever more.

MEASURING TIME: THE EARLY DAYS AT BEACON

In its early days, Beacon Charter High School for the arts, was located at 35 George Street. The building was bordered by Rathbun, Privilege, and Social Streets, and situated in the former St. Louis Parochial School, approximately

Table 7.1. The Number of Days in a School Year for Some Major US Cities (1842 and 1892)[a]

	Year	
City	*1842*	*1892*
New York	245	203
Chicago	240	192
Philadelphia	252	201
Detroit	259	196
Buffalo	240	195

[a]Adapted from Gold (2002)

one mile from its present location at 320 Main Street. The 100-year-old, three-story brick veneer structure (originally a two-story wooden building in 1910) was used by Beacon during the week as a public school. On Sundays it was used for religious classes for the children of All Saints Catholic Parish. Although it was primarily occupied as a public school, no one who entered the building could help but sense they were on holy ground. Very often crosses and children's Sunday-school artwork were left in the classrooms. Each Monday, Beacon teachers would have to conceal the religious materials. In the main office, a five-foot statute of Jesus had to be rotated to face the wall in case we had visitors from RIDE (Rhode Island Department of Education). The main office was located on the first floor mid-way down the hallway and managed by the school's only secretary/receptionist.

Besides normal secretarial duties, her major task was to signal the start of the school day, the times at which classes ended, and the end of the school day. On her desk was a 60-minute egg timer. Unfortunately, however, classes were 90 minutes long. Thus, her procedure for signaling the end of class periods involved setting the egg timer for a full 60 minutes and then resetting it for the remaining 30. After these 90 minutes had passed, she would get up from her desk, walk out of the office into the hallway, and push a button on the wall. It rang a bell and so indicated the change of class periods. As is the case with most school secretarial positions, the often agitated interaction with students, parents, and staff, the distraction-to-productivity ratio in her office was high. Remembering to reset the timer was problematic. Very often, the bell was rung 10 or 15 minutes late, or not at all. And, the result was then confusion. Classes were dismissed at diverse and disparate times because teachers assumed the bell had been forgotten. Of course, under this system, a strategy for selfishly extending the lunch period was to initiate a distraction down at the main office by making a strategically timed phone call to the secretary and posing a cleverly crafted question to her.

It was not long before Principal Pilkington made a decision to abolish all bells and had clocks installed in every classroom. Every day he would ask teachers to synchronize their clocks with the main office. This "no-bell" policy continues to this day—definitely a good thing because no one can find the egg timer and the budget is still tight.

MEASURING TIME: MRS. WATERS' CHEMISTRY BLOCK 1 CLASS

> Time is water in the desert, a teacher's most precious resource, to be husbanded, guarded, and conserved. Every minute matters.
>
> —Doug Lemov, Teach Like a Champion

During the very first few days I spent in Mrs. Waters' Chemistry Block 1 class, it quickly became apparent that, for her, time was a precious commodity. Class started on time, activities and learning objectives for the day were written on the board at the start of the session, copies of handout materials had been prepared. Overall, there was little or no loss of instructional time (to be defined later) in the transition from one activity to the next. Mrs. Waters was a "bell-to-bell" teacher, even if the bells were silent. She had paced her curriculum for the semester in keeping with the state science standards that were to be addressed prior to the May testing dates. However, as the days progressed, interruptions to her well-choreographed pedagogical program became increasingly apparent. I began keeping track of lost time in the classroom due to teacher and student absenteeism, field trips, snow days, student behavior, phone calls to the classroom, etc. These interruptions were not unique to this class, nor were they new to me. It was just that I had never tried to quantify this in my own classroom. I was now entrenched in four-dimensional space-time. Events were simultaneously being experienced from both a teacher's and a student's perspective, and it was becoming ever clearer to me just what a precious commodity classroom time really is.

Lost Opportunities to Learn in Mrs. Waters' Chemistry Block 1 Class

The 180 school-day calendar sets a limit on the number of chemistry classes that can be held. In Rhode Island, no state mandates dictate the number of days per week on which a particular subject must be taught. The length of time for each class is defined by each school district. In our case, Beacon's administration decides the schedule. Since we are an art-focused school, scheduling art classes takes priority. This is clearly noted in Beacon's original Mission Statement:

> Beacon's Mission Statement
>
> Beacon Charter High School for the Arts prepares young student-artists for advancement through secondary and post-secondary education while immersing each student in the art discipline of their choice and providing them with the practical business experience needed to be successful in the sustained practice of their art.

The key phrase in the school's mission statement is "immersing each student in the art discipline. . . ." Every student at Beacon takes an art class each day for the full 180 days of the school year. To compensate for this, academic classes such as science, math, English, history, and foreign language are limited to 90 school days. We immerse our students in art to such a degree that

I sometimes think they should be issued snorkels to help breathe in their academic classes. Thus, as concerns chemistry class, Mrs. Waters theoretically had 90 days at most with which to work. Unfortunately, reality often plays tricks on us. In what follows, I begin with a macroscopic view of the reality of these 90 days and then narrow into a microscopic one, a look at the class from within and as a student.

A Macroscopic View: Days Lost

February 4, 2011 (Day 1) Friday—Teacher PD. Today should have been the first day of chemistry class. However, since "Study Island," a web-based tutorial and assessment tool, is going to be an important part of the chemistry curriculum for preparation for state testing, Mrs. Waters, Mrs. Taylor (Special Education Director), and myself are participating in a webinar during Block 1 (the scheduled time for our class). Also joining us is Ms. Perkins, a math teacher who will be using the program for her classes. Mr. Connell, a special needs teacher, has been assigned to cover Mrs. Waters' class. In calculating the lost days, I consider any chemistry class without a certified science teacher as instructor a lost instructional day—even if work has been deposited in advance with the substitute teacher. This therefore is a lost day as far as instruction is concerned.

February 10, 2011 (Day 5) Thursday—Mrs. Waters is out. I arrived at class today to find that she is absent because of a sick child. Mrs. Oswik, the school's Health and Nutrition Advisor, is covering our class and is giving a "sex education" talk. Fortunately, both Mrs. Pagliaro, the Microsoft Office and business teacher, and Miss. Wall, a humanities teacher, have requested my services with SAT math preparation in their senior classes—the perfect excuse to escape Mrs. O's sex talk. So, today is recorded as lost.

February 18, 2011 (Day 11) Friday—Juniors take the ASVAB. Today all juniors are taking the ASVAB (Armed Services Vocational Aptitude Battery) in the cafeteria. It is a multi-aptitude test developed and maintained by the Department of Defense. It tests students' knowledge in four critical areas—arithmetical reasoning, word knowledge, paragraph comprehension and mathematical knowledge and is used by the military to qualify applicants for certain occupations. Beacon uses the test as a means of helping students identify their strengths and weaknesses and to identify any areas of potential occupational interest for which they may qualify. The test takes about 2-½ to 3 hours. Again, today is recorded as lost where chemistry instruction is concerned.

February 28, 2011 (Day 12) Monday; and, February 29, 2011 (Day 13) Tuesday—PD for teachers, field trips for students. Field trips are scheduled for students on these days. Only essential faculty will remain behind for PD (professional development) on the new Common Core State Standards. So, now we have lost two more instructional days.

March 4, 2011 (Day 16) Friday—Mrs. Waters is out. When I arrived in class today, Ms. Perkins was sitting in front of the class. She said something to the effect that Mrs. Waters' daughter was sick at the daycare. Since this was an emergency situation, she had not left any instructions for the lesson. I was, however, able to do a fast dance and get the mobile computer cart so the class could do a Study Island assignment, but nevertheless recorded today as a lost instructional day. Several days later I asked Mrs. Waters what happened. She explained that her daughter Madi had gotten sick and thrown up just as they were arriving at the daycare at 7:10 a.m. She had to take her to the doctor. The doctor sent Mrs. Waters to Urgent Care for an X-ray. But she ended up having to make two trips there—a second one with a neighbor who could help with Madi. This took several hours so there was no way she could teach that day.

April 5, 2011 (Day 38) Tuesday; and, April 7, 2011 (Day 40) Thursday—Juniors piloting the NECAP Writing Test. On April 4, 2011, 3:33:18 pm EDT, an email is sent out to faculty. It seems that someone has volunteered our juniors to pilot the software for a new computer-based NECAP writing assessment. The email states "this is just a pilot, nothing will be graded, just to test out the software, etc." The juniors had already undergone three stressful days of the state's NECAP assessments in reading, math, and writing last October and will be taking three days of the state's NECAP science assessments five weeks from now in May. Still more opportunities for instruction are being lost! This email messages also contains the schedule for the testing that is to be held on Tuesday and Thursday this week. It looks like Mrs. Waters' chemistry classes will be affected both days. Unfortunately, Thursdays are her lab day and so she will have to make a further schedule change. Given that the time needed for preparation of her students for the NECAP science assessments was so precious, I asked Mike Skeldon why the chemistry class period was being used for the software testing. He explained that the state required that at least 50% of the juniors be tested at the same time, and then said that he had tried to find a schedule that was least disruptive where all of the teachers were concerned.

As it turned out, Mrs. Waters' Chemistry Block 1 class was not affected by the writing pilot, but her other two chemistry classes were. Therefore I did not consider these instructional days as lost.

May 2, 2011 (Day 52) Friday—Mrs. Waters out sick. The class is assigned to the computer lab to work on Study Island because the "Birds of Prey" lady (a guest lecturer who brings live killer birds with her so that the visual arts students can sketch them) is using Mrs. Waters' classroom. The assignment was to work on a Study Island topic related to the student's weakest NECAP area. So, today was also recorded as a lost instructional day. A few days later, I asked Mrs. Waters why she was out. Apparently, her family decided to go to a Red Sox game on Sunday—the weather was beautiful. On the way home, the kids started getting sick as did she and her husband, Chris. Later at home, and into the night, things got still worse. So, on Monday, she had to stay home and take care of the family. Even Chris was confined to the house in spite of the fact that he never misses work and had not been this sick in the twelve years since she had known him.

May 13, 2011 (Day 61) Friday—Mrs. Waters out sick (Food poisoning). May 13, 2011 5:01 am EDT, it's just four days until the NECAP science test. I am half-asleep, and awake to the sound of my cell phone ringing near my pillow. I am receiving the following text message from Mrs. Waters:

> John, I need you to help me out today... I haven't even gone to bed yet. I am sick, fever, chills, vomiting. I have been trying to fight it but I can't even stand up. Chris [her husband] has to take me to the emergency room. Can you please have Block 1 finish the periodic table thing and I'll send you the 'problem of the day.' Block 2 and Block 3 need to finish going over the 'Location' inquiry task. Maybe Shayna can help. Everything is in the mailboxes. I am beating myself up about this!! I don't think Mike is in, that's why I'm texting you.

5:03 am, my response:

> "No problem. I'll take care of everything—relax this weekend."

Mrs. Waters texts back:

> "Thanks so much. I really appreciate it."

I was able to prepare the work for Block 1 as Mrs. Waters requested but recorded today as a lost instructional day as well.

Later that day, at 3:02 p.m. I receive a text from Mrs. Waters:

> Just got out of the ER. I was extremely dehydrated and had to have three bags of fluid. Very bad case of food poisoning. I feel like I'm letting everyone down for NECAP.

My response:

> "You're not letting anyone down. In addition to your instructions, I gave everyone three PODs. Call me when you feel better."

May 17, 2011 (Day 63) Tuesday—First day for the NECAP science assessment; May 18, 2011 (Day 64) Tuesday—Second day for the NECAP science assessment; May 19, 2011 (Day 65) Tuesday—Third and final day for the NECAP science assessment. All eleventh grade students took the Rhode Island NECAP science assessments. They were administered over a three-day period. Students did not have chemistry class on these days and therefore three additional lost instructional days.

June 10, 2011 (Day 80) Friday—Final Exam day for Block 3 and Block 4. Mrs. Waters' Chemistry Block 1 and 2 classes had their final exam yesterday so, today counts as a lost instructional day, too.

All juniors were required to start school two weeks earlier in the Fall of 2010 for preparation for the NECAP October assessments in the areas of mathematics, reading, and writing. We refer to this as NECAMP. As an incentive, they were promised that if they raised the school's scores and helped us meet the state's AYP (adequate yearly progress) targets for Beacon Charter High School, they would get out earlier at the end of the year. In February of 2011, the scores were made public (we exceeded the state average in every category and made all of our targets) and, the juniors were informed that they would be released 10 days earlier as promised. Since they had no chemistry instruction during the two weeks of NECAMP, nor during the two weeks that were cut from the end of the year, I recorded an additional ten lost instructional days. (June 13, 2011 (Day 81) Monday; through June 24, 2011 (Day 90) Friday).

So, we were confronting the reality that of the 90 days theoretically allotted for chemistry, 22 were lost and we were left with 68 actual instructional days. Checking school records, I found that students in my Chemistry Block 1 class were absent on average four days over the semester. Working attendance into my calculation, therefore I concluded that chemistry students actually had an average of 64 instructional days out of the 90 theoretically allotted to them. But it gets worse!

A Microscopic View: Minutes Lost

Rhode Island law specifies that the required length of a school day for elementary and secondary schools must contain not less than 330 minutes of actual school work. This excludes lunch, recess periods, common planning time, pre- and post-class time for teachers and any other time that is not actual instructional time. An analysis of Beacon's schedule revealed that the school does meet this requirement. However, the state does not specify the length of any given class. For Chemistry Block 1, the average length across the five days we met each week was 78 minutes.

Using my calculations of 64 instructional days and 78 minutes per class, there are 4,992 minutes available to Mrs. Waters for Chemistry Block 1 instruction. This may seem like a lot of time, but consider that there are theoretically180 days in the academic year and a total of 330 minutes of state mandated instructional time per day, i.e., there are 59,400 total instructional minutes per school year. So, by my calculation, 4,992 minutes out of the 59,400 total equates to about 8.4% of the students' school year. This was the amount of time dedicated to science, specifically, to instruction in Chemistry Block 1. That is less than 10 percent of the total! To appreciate this fact, try going to the *Dollar Store* with just a dime in your pocket—see how far it gets you!

But, we haven't yet taken in to account all of the interruptions that occur within the 78-minute class period. These, of course, further diminish the amount of time theoretically available.

And, there were, as noted above, many interruptions to classroom instruction during my 79 days. I define an "interruption" within class time as any unanticipated event that caused Mrs. Waters to pause, stop, change direction, etc. Cases in point would be an announcement over the PA system, a fire drill, the arrival of unexpected visitors, a technology problem, a student needing to leave the classroom and the need to fill out a hall pass for this, a telephone call, unsolicited student questions, students arriving in class late, and various disciplinary issues. Some of these are minimally distracting, others eat up significant amounts of valuable time. No matter what their duration, they decrease the amount of instructional time available for a class. As my research was of an exploratory nature, I did not try to document every instance of an interruption—I leave that to some future researcher. However, when I did observe an incident that amounted to more than a short interruption—beyond a minute or two—I recorded it in my journal. I offer the following examples of such incidents.

February 17, 2011 (Day 9) Thursday—Mrs. Waters arrives late due to a meeting with RIDE. Both she and myself were 15 minutes late to class today because we were being interviewed by a team from RIDE on our participation in the pilot of the new Rhode Island Model Educator Evaluation System.

March 15, 2011 (Day 23) Tuesday–PA Announcement and technology problems. According to what was written on the board today, we would be doing a "Problem of the Day" and then a research activity in groups. Mrs. Waters explained that the latter was preparation for a test in a few days. After completing the "Problem of the Day," she began assigning partners for the group activity. Just before she finished, a voice came over the PA: "Please excuse this interruption, but any junior or senior interested in meeting with representatives of the Art Institute, please report to the cafeteria." At first, nobody expressed an interest in leaving. Mrs. Waters continued to explain the group activity. Each group was assigned one of six scientists who had developed models of atomic structure. We were to research the major features of the model and create a poster of our findings. The scientists were John Dalton, JJ Thompson, Ernest Rutherford, Niels Bohr, Louis de Broglie, and Ernest Schrodinger. Then the groups were to position their posters in chronological order on the wall so that all students could take notes from them. This is a type of "jigsaw" learning strategy. It allows coverage of more material in a shorter period of time and promotes student-centered learning and engagement.

Suddenly, Ruby interrupts Mrs. Waters and says that she and Katy want to leave class and go to the Art Institute talk in the cafeteria. Mrs. Waters again asks if anyone else wants to do this. Two other students also decide to go. This interruption now affects the whole class, since two of the study groups have now lost members. Simultaneously, computer problems developed. They were related to wireless internet access and were hampering our research on these scientists. Today was not an efficient instructional day through no fault of Mrs. Waters. This activity had to be completed in the following class. I made a note in my journal to recommend that the administration inform teachers in advance when colleges will be on campus. Students should be required to sign up for such events ahead of time so that this type of interruption could be avoided. A few days later, between classes, I ran into Nichole, a Chemistry Block 1 classmate. During our conversation, Nichole blamed Ruby for the fact that the class had not finished the poster activity.

March 18, 2011 (Day 26) Friday—Ruby's protégé fills a void. Mrs. Waters had once mentioned to me that this would be an ideal class if Ruby wasn't so disruptive. Today should have been a nice teaching and learning day given that she was absent. Unexpectedly, however, Jasmine [pseudonym], who normally sits next to Ruby, became disruptive for much of the class and Mrs. Waters had to deal with this at several points.

March 23, 2011 (Day 29) Wednesday—State bureaucracy at work; Ruby; wireless problems. Today is Study Island day, but Mrs. Waters has been informed that she has been asked to have all students in her class do an online survey for RIDE (Rhode Island Department of Education). On the board she has written:

> "First complete survey at http:/tiny.cc/RIhighsurvey"

Completing the survey relieved us of about 10 minutes of instructional time.

Next, there was an incident with Ruby and Jasmine. Jasmine suddenly left the class visibly upset. A few minutes later, Mrs. Murphy, Dean of Students entered the classroom and questioned Ruby. She responded by being obnoxious and rude. Later, Mrs. Waters' left the classroom for a few minutes to check on Jasmine.

Also, I noted that there was a significant amount of downtime on the Study Island activity today due to further wireless internet access problems.

Near the end of class, Ruby was told by Mrs. Waters to stop talking. But she continued and all of a sudden Siraj looked at her and scolded her: "What don't you understand about being quiet?" Before she could respond, Mrs. Waters proactively tells Ruby, "Don't say a word!" The class ends with no further incident. Still, all of these interruptions ate up valuable instructional time.

March 29, 2011 (Day 33) Tuesday—Ashley walks out. Mrs. Waters was going over her PowerPoint presentation entitled the "Structure of Atoms Unit" when suddenly Ashley jumped out of her seat and announced to the class that she has ADD (Attention Deficit Disorder)—as if we didn't already know this. Then she left the classroom without asking permission. Mrs. Waters stopped her instruction to call Mrs. Murphy, the Dean of Students. Another interruption that causes Mrs. Waters to have to get the class back on track.

March 31, 2011 (Day 35) Thursday–Culinary students missing in action. Just as we were arriving in class today, a voice came over the PA: "Please excuse this interruption, would all junior culinary students please report to the kitchen." Immediately, Manar, Sarah, Sima, and Siraj take off for the kitchen, leaving their belongings by their seats. Mrs. Waters and I are thinking that perhaps a mess had been left behind in the kitchen and Chef was calling his students back to clean it up.

Today, Mrs. Waters had reserved one of the fourth-floor computer labs for our class to work on our formal lab reports. She proceeded to walk with the class up to the lab. I told her that I would stay behind until the girls got back so they could get the things they left behind. After 20 minutes had passed, I began to suspect that the girls, all extremely well-behaved, conscientious, and responsible, perhaps weren't coming back anytime soon and that I had better go investigate.

When I got to the kitchen, I found Brittany, Lindsay, Manar, Sarah, Sima, and Siraj, all from our chemistry class, hard at work preparing food for the senior Capstone presentations that were going on upstairs. They told me they didn't know they would be tied up all period. I told them I would lock their things up in the classroom for safekeeping. They could search out Mrs. Waters or myself when they were done. I proceeded upstairs to the lab to inform her what was going on. She said she knew that Brittany and Lindsay would be out because they had asked permission ahead of time, but she was never asked or informed about the schedule changes for the other four students.

April 14, 2011 (Day 45) Thursday–reduced time 11:30 a.m.–12:20 p.m. ("Clash of the Classes"). Today's class length was reduced from 78 minutes to 50 to allow for a school-wide event over at the CYO (Catholic Youth Organization) recreation center just one block from Beacon. Students formed teams and competed in various events.

OTHER OBSERVED INTERRUPTIONS

"Mrs. Waters, your phone is ringing!"

I think one of the most aggravating interruptions for a teacher during a class is receiving phone calls—especially calls that could have been made between classes or after school. Mrs. Waters had more than her fair share of calls over the many days I spent in class.

Adding to the significance of all of these interruptions was the fact that her phone was inconveniently located by the door, about 17 feet from her desk and 24 feet from her normal teaching position in the front of the class. So, time was needed for her to move to and from the phone. Sometimes she would have students answer it—but she still had to get involved to some extent.

There is no doubt in my mind that phone calls often changed the pace and tenor of class activities. I mentioned this to Pat Hawkridge, the school's Artistic Director, and she agreed. She had just been in an arts class in which she had counted eight incoming phone calls.

"Mrs. Waters, can I go to the bathroom?"

Sometimes it seems to me that every classroom should be outfitted with a PSU (portable sanitation unit) to cut down on the number of interruptions for bathroom breaks and the corresponding loss of instructional time. Beacon, like most schools, requires students to have a signed hallway pass when they exit the classroom. This means that the teacher has to stop what they are doing and sign the pass.

"Where's Ashley?"

On Wednesday, March 30 (Day 34), I decide to do a mini-study on student engagement in a chemistry activity. Every few minutes, I recorded an "e" next to the respective student's name on a class roster if that student was engaged. When I got to Ashley's name, I noticed that she was out of her seat—in fact out of the classroom altogether. Simultaneously, and without having been given any notice by me, Mrs. Waters' realized that she has been gone for about 15 minutes (initially she had been given permission to go to the bathroom). She started to head for the phone to call the main office, when Ashley suddenly walked in. So, we now had still another interruption to the process of instruction.

SOME POSITIVE OUTCOMES

As concerns "time," the subject of this chapter, there is little that a teacher can do about the limitation of 180 school days, the fixed number chemistry classes, and the 78-minute class length. However, school policies and classroom management are two areas where efficiency with regard to time can be practiced and refined.

During the course of writing this book, I made several simple suggestions to Beacon's administration based on these findings. Below are four of them. These were implemented for the 2011-2012 school year and produced positive results.

1. *Fewer PA announcements*: The administration significantly reduced the number of PA interruptions and is now more consistent in the timing of daily announcements in the morning and afternoon.
2. *Fewer phone calls allowed:* There is now a greater school-wide awareness that phone calls to classroom teachers during class time constitute a disruption and so should only be made in cases of emergency.
3. *Students must sign up in advance for college visits*: College visits are announced in advance, so students must sign up for them in advance. This allows teachers to plan for their absences on certain days.
4. *Sign-out/Sign-in:* When they are permitted to leave the classroom, students should mark their student planner with time of departure and destination. This will substitute for their hallway pass. The teacher does not sign this at this time, however. Students then fill out a sign-out/sign-in sheet that is placed on a table by the door. At the end of class, the teacher initials the student planner thereby verifying that that student had followed school policy for leaving the classroom. One consequence of this formal procedure has been that fewer students have left the classroom and they have remained away for shorter durations. Clearly, they are aware that we are tracking them. Further, with regard to RTI (Response-to-Intervention) initiatives, we now have a more efficient means of looking for patterns among certain students, i.e., those who continually ask permission to leave the classroom.

A TAXONOMY OF CLASSROOM TIME: ESTABLISHING DEFINITIONS

As part of my exploratory study in Mrs. Waters' classroom, I observed and recorded events relative to the use of "time." Many of these have now been discussed in detail. In conjunction with this study, I investigated the research literature in an effort to adopt and maintain existing terminology and definitions wherever possible when it came to classifying time. Definitions are provided in table 7.2. There is some subtle variation among them, but these are generally accepted and comprise my taxonomy of classroom time.

Of the 16 time-related terms contained in table 7.2, "school year" may be considered the "whole" and the others the "fractional parts." As concerns the "whole," the teaching and learning process is framed by the six major time components shown in figure 7.2. The outcome of this process is *academic*

Table 7.2. A Taxonomy of Classroom Time: Definitions

Time Related Term(s)	*Definition/description*
School Year	The minimum number of days that public schools are required to be open by state law. Per R.I.G.L. 16-2-2, a minimum of 180 school days annually, exclusive of holidays, is required for students.
School Attendance	The number of days a student actually attends school.
School Day Length	The number of hours in a school day. By Rhode Island law, secondary schools must have no less than 5 ½ hours (330 minutes) of actual school work, excluding lunch, recess periods, common planning time, pre and post teacher time and any other time that is not actual instructional time.
Allocated class time	Allocated time is the theoretical scheduled time planned by the state, district, or school for instruction of a given subject (Berliner, 1990). For Beacon Charter High School, this was determined to be 78 minutes for 90 days or 117 hours per school year.
Class Attendance	The number of classes of a given subject that a student attended. It can never be greater than the school days attended but can be less.
Instructional time	The actual time used within the allocated class time by the teacher for instruction of content related material. I define this as the time used by the teacher to actively demonstrate, inform, instruct, lecture, point out, etc.
Transition Time	"... usually defined as the noninstructional time before and after some instructional activity." This includes time used for taking attendance, collecting homework, passing back papers, putting books away, cleaning up after a lab or art activity, etc. (Berliner, 1990, p.5). See also Lemov (2010).
Waiting Time for Help	"... usually defined as the time that a student must wait to receive some instructional help" (Berliner, 1990, p.6).
Unannounced/ Unexpected Interruption Time	I define this time as interruptions during class, not anticipated, that cause a teacher to pause, stop, change direction, etc. in instruction. Examples would include disciplinary issues, unannounced fire drills, phone calls, students needing to leave the class, and technology problems.

Time Related Term(s)	*Definition/description*
Announced/Expected Interruption Time	I define this as interruptions during class, that are anticipated but cause a teacher to pause, stop, change direction, etc. in instruction. Examples would include announced visitations, field trips outside the content area, and announced fire drills.
Engaged Time	"... usually defined as the time that students appear to be paying attention to materials or presentations that have instructional goals...A synonym for engaged time is 'attention'" Berliner (1990, pp.4-5). It implies a voluntary mental attitude of concentration. Engaged time is not synonymous with *time-on-task* (Danielson, 1996).
Time-on-task	This term is often interchanged with the term *engaged time* when they are not synonymous. "The term *time-on-task* has a more restricted and more complex meaning than does the term *engaged time* ... a student is deeply involved in mathematics or a comic book during a time period allocated to science" (Berliner, 1990, p.5) and thus is not on task.
Academic Learning Time (ALT)	"... usually defined as that part of allocated time in a subject-matter area (physical education, science, or mathematics, for example) in which a student is engaged successfully in the activities or with the materials to which he or she is exposed, and in which those activities and materials are related to educational outcomes that are valued (Berliner, 1987; Fisher et al., 1980) ... Academic learning time is often and inappropriately used as a synonym for engagement, time-on-task, or some other time-based concept" (Berliner, 1990, p.5).
Aptitude	"... usually defined as the amount of time that a student needs, under optimal instructional conditions, to reach some criterion of learning. High aptitude for learning something is determined by fast learning; low aptitude is reflected in slow learning" (Berliner, 1990, p.6).
Perseverance	"... usually defined as the amount of time a student is willing to spend on learning a task or unit of instruction" (Berliner, 1990, p.6).
Pace	"... usually defined as the amount of content covered during some time period" (Berliner, 1990, p.6). Mrs. Waters' teacher's edition of her chemistry book, with 25 chapters, had a pacing guide based on 36 weeks (180 days).

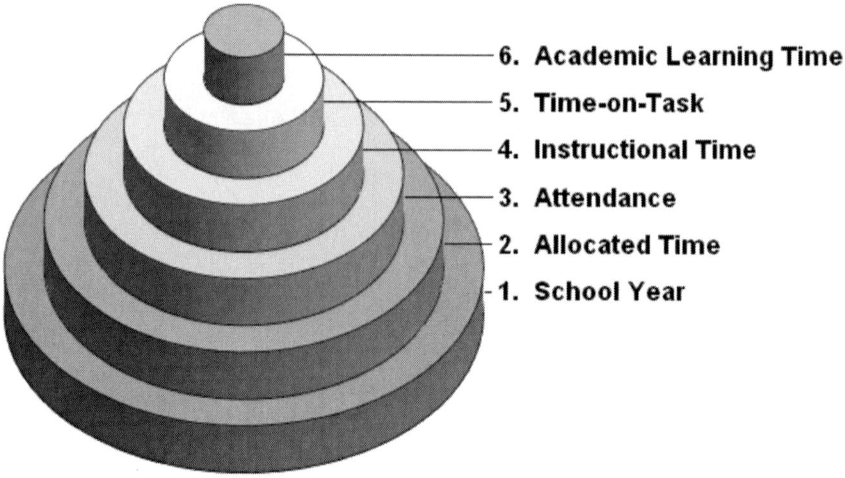

Figure 7.2. Taxonomy of Time—Major components.

learning time (ALT)—the actual time a student is successfully engaged in learning and acquiring information of value. As Berliner (1990, 18) states: "It is recognized that even time-on-task is not quite the most desirable measure to study about a student's behavior. What is really wanted is a measure of time-on-the-right-tasks." The goal is to maximize ALT. This requires a well-choreographed approach to classroom management with regards to the taxonomy of time. (see figure 7.2).

In his highly acclaimed book, *Teach like a Champion*, author Doug Lemov states in reference to this problem of appropriate use of classroom time:

> You can read all day long and well into the night about the need for more resources in public education but every teacher has, at his or her fingertips, an opportunity to increase the scarcest and most important resource that money buys by a significant amount (p. 154).

According to Lemov, identifying the components of time such as transition time (see definition, table 2) is a prime target in analysis. He states that if a teacher were able to cut one minute off of each of 10 transitions made per day and sustain this over the school year, nearly a full extra week of instructional time could be gained. Further, Lemov points out that messy transitions are an invitation for students to display their behavioral problems. He advocates for "quick and routine transitions that students can execute without extensive narration by the teacher—that is, Tight Transitions …" (p. 154).

Lemov claims he often begins teacher trainings by showing a clip of his colleague, Doug McCurry, founder of Amistad Academy in New Haven Connecticut. McCurry is spending time in the first two days of the school year teaching students to pass papers back efficiently. Just even asking them to perform this simple routine throughout the school year yields several days worth of added instructional time.

He also discusses the minutes that are lost at the end of class. All too often teachers write these minutes off, thinking that it is too late to start something new or assuming that quitting early will be regarded as a reward for hard work. He calculates that for a teacher who teaches six classes per day over a school year, four wasted minutes at the end of each class equate to over seventy hours, i.e., several weeks, of instruction time.

MORE "BEST" PRACTICES FROM MRS. WATERS' CLASSROOM WHERE TIME-SAVING IS CONCERNED

One great time-saving strategy employed by Mrs. Waters was the use of the "Pre-Lab Activity" worksheet. It was assigned as homework the night prior to each lab experiment. It involved reading about the experiment that would be performed in class the next day, noting relevant safety concerns, and flowcharting the procedure. Students were required to sign the worksheet acknowledging that they had completed the reading and understood what needed to be done. This made it possible to allocate more valuable classroom instructional time to experiments. This strategy falls within the paradigm known as *flip teaching* or *flipping the classroom* whereby students study the topic themselves the night before class. The next day, the students apply their knowledge in problem-solving activities with the teacher in the role of facilitator/tutor. Research over the last decade has shown that flip teaching increases student engagement, students learn more, more classroom time is made available for the teacher to work with the neediest, and it decreases the dropout rate.

One of the most oft quoted adages in America's factories and businesses has been "Time is money." Still, we have never heard it within the walls of school buildings! Early in my twenty-plus year career as a mechanical engineer who designed production machines and developed manufacturing processes, I learned to appreciate several simple facts: The shortest distance between two points is a straight line, efficiency in processes generally translates into profit, and getting to a 95% solution is easy, it's the remaining 5% that often requires the greatest effort.

How Does This Relate to Mrs. Waters' Class?

As I sat in Mrs. Waters' class, I noted several items in my journal. These were opportunities for increasing efficiency in the teaching and learning process relative to the time allotted for this. The common thread between all of these seemed to be organization. Some of them simply involved physical changes to the classroom, others, improvements to instructional materials and pedagogy. My premise is that saving minutes on routine activities and sustaining these newly acquired forms of efficiency over the longer term (e.g., the school year) can free up hours, days, and even weeks of additional instructional time. The following are some of the time-saving practices that were put in place as a result of my being a participant-observer in Mrs. Waters' class.

Physical Changes

Mrs. Waters had arranged her classroom tables for student-seating in a "U" shaped configuration with the open end towards the whiteboard. Students sat two to a table along the outside of the "U." Although this arrangement was different from the traditional rows and columns of desks typical of most classrooms, it was ideal for Mrs. Waters' style of teaching. It provided everyone with a full, unobstructed view of the whiteboard and also gave her an unobstructed view of all her student's faces. It should have allowed her to navigate 360 degrees around the "U" and so ensure easy access to each individual. However, on the outside of the "U" where I sat, and behind my seat, there was a long table that was used for storing lab equipment. It was positioned immediately up against an outside wall such that no further walking space could be gained. Therefore those of us who were seated at our tables on this side inadvertently prevented anyone from walking behind us. Mrs. Waters was therefore also blocked from circling around the outside of the "U." Further, I found myself having to get out of my seat several times during class so that fellow students could come in or go out. After the first few weeks, they learned to walk the long way around the "U" rather than inconvenience another student by forcing them to get out of their seat. So, I concluded that the table behind me, which was heavily loaded with lab equipment, had to go. This would mean finding a place to store all of the lab equipment.

Louis Pasteur's famous quote "Chance favors the prepared mind" comes into play here again. I happened to recall that, in December 2010, Beacon had purchased some items that had been auctioned off by a local middle school. These had been placed in storage in a back room on the fourth floor of our school. I remembered seeing some large metal cabinets and wondered if anyone had claimed them. I quickly made my way to the storage area. There were four locking, double-door heat-resistant storage cabinets—in all, about a $2800 value new. They had been purchased for a total of $100 and were in great shape except for some minor dents, mismatched colors, and a little dirt. After measuring them and checking the available space in Mrs. Waters' classroom, I decided to take on the project of cleaning them up, knocking out the few minor dents they contained, and repainting all of them in a glossy black. This ended up taking about 40 hours of my time (most of it done, as noted, during our April school vacation week). Still, they worked out perfectly for the chemistry classroom. I was able to store every piece of lab equipment including test tubes and other glassware, triple-beam scales, burners, safety glasses, etc. in them. All four cabinets fit perfectly up against one side of the classroom; and, when they were stocked with lab equipment, they freed up lots of space in the classroom. Most importantly, they made it possible to remove the table behind my seat that was obstructing the flow of traffic. The day after these were put into place, Mrs. Waters was able to move with ease around the outside of the "U."

Friday, April 22, 2011 was the last day of the school vacation week. I happened to be in the building to do some painting on the storage cabinets when I decided to check to see what other gems may be hidden in amongst the auction items. And, I struck gold! Buried amongst some miscellaneous furniture was a strange looking bookshelf. It did not have any shelves, but did have a lot of protruding pegs that looked like they had been inserted for positioning shelves. I located a box of 28 wooden square pieces not too far away and then realized what I had acquired. It was a mail slot cabinet with four columns—each with seven shelves for a total of 28 pigeon holes. This would be perfect for Mrs. Waters' class as she had worksheets for class activities almost every day. What a great organizational tool! Mrs. Waters loved it and quickly made good use of it. It was a real time saver, too, since she did not have to search through piles of papers. And, students could easily be directed to collect any of the papers they needed because they were stored in the mailbox. Mrs. Waters' time was therefore freed up.

Some Improvements to Instructional Materials and Pedagogy

As noted, Mrs. Waters had an activity at the very beginning of class. There was a POD (Problem of the Day) and very often, a COD (Chemical of the Day). Students were instructed to begin working on these tasks as soon as they entered the classroom. The use of the acronyms POD and COD were my suggestion. After the very first few days of the semester, I had found it tedious to keep writing out the entire phrases. In a very short time everyone in the class was speaking and writing the acronyms. This may seem trivial, but tedious/repetitive tasks often lead to boredom and disengagement.

The POD for chemistry class was generally a practice problem for the state's NECAP science assessments. Mrs. Waters would either provide a problem statement on the board, or project it onto the whiteboard.

The COD was a chemical compound that students were being introduced to. Mrs. Waters would write the chemical name, the molecular formulae, structural formulae, characteristics, note where it is used, and other facts on the board. Students would then copy the information into their notebooks for reference.

Very often, the whiteboard would be used for instruction after the initial activity. This meant that the COD and/or POD (if written, as opposed to projected) would have to be erased because writing space was needed. The result was that Mrs. Waters had to regularly rewrite the problems for each of her two other chemistry classes. Projecting the COD and/or POD proved to be the most efficient use of time.

As a means of saving time and improving understanding of the information being copied from the board, I developed a standard template for the COD in the form of a graphic organizer, and also a standard format for the POD. These were made available for the students at the beginning of each class and provided several benefits, among them: reduced time for student writing and so increased instructional time, improved student notebook organization, better understanding and retention of the content being taught; it also provided an improved student resource for quiz and test preparation.

On February 14, 2011 (Day 7), Mrs. Waters came around our class to check the homework assignment. Part of it had been to work on PowerPoint slides. I had completed them. An additional assignment required that we look up "green house gasses." I had, however, forgotten to write that one into my notebook. I vowed that this would never happen again! That night I created a "Homework Log Sheet"—a general template for recording assignments and their respective due dates. I gave it to Mrs. Waters to distribute to all of her chemistry students so that they could keep it in their notebooks. This would hopefully minimize their forgetting assignments like I did, and it could cut down on the time required for her to track missing assignments.

This had a pleasant outcome—about two weeks later, I heard Mrs. Waters tell Nichole to "go show John." Nichole had designed a "Chemistry Extra Credit Graphic Organizer" based on my homework log to record the "extra-credit" opportunities that Mrs. Waters often provided. Nichole shared her log sheet with others in the class. This bit of effort at improving organization ended up being a major time saver for everyone. Mrs. Waters had previously taken class time to grade, and record receipt of, extra credit work. Often, it had been done by several members of the class. But now she was able to change her process. Each student was responsible for handing in the log sheet at the end of the grading quarter so that Mrs. Waters could quickly run through it and record all of the grades in a single interval.

In their article, entitled "The 'I' in RTI: Research-based factors for intensifying instruction," authors Harlacher, Nelson Walker, and Sanford (2010) recommend that teachers solicit *unison responses*. "Using unison oral responding instead of calling on one student at a time can increase a student's opportunity to practice by six times in a group of six students, thereby increasing engagement and successful practice" (p. 35). Although I agree with the authors, I would suggest a quieter but equally effective strategy: As a means of attaining high levels of student involvement and efficient use of instructional time, have all students in the class occasionally respond to questions or problems by simultaneously holding up their answers on a piece of paper or personal mini-whiteboard. For example, in a class of 20, the teacher asks every student one question at a time (very common practice) and doing this (asking and answering the question) takes an average of one minute per person. It will therefore take 20 minutes for every student to have participated; 20 responses will be forthcoming. By contrast to this, consider what would happen if every student had their own personal whiteboard. All of them respond to each question by writing their answer on it and then hold it up when the whole class is prompted. In twenty minutes, the teacher will have asked twenty questions but 400 student responses will have been forthcoming—quite a significant increase in opportunities for engaging students and a real time saver. It makes it much easier to assess who is "getting it" and who is not. I suggested this strategy to Mrs. Waters at the end of the semester, and she said she would try it in the future. As it turns out, the future came relatively quickly and with technology that went well beyond what I was suggesting. Just a few months later, I found Mrs. Waters experimenting in her classes with texting devices in conjunction with online programs. These can analyze unison responses from students and display the results visually, e.g., bar graphs, charts, and word clouds. Such technology most definitely makes more efficient use of time. I predict that student response technology will be a widely used tool in the classroom of the future.

TIME TO STOP TALKING ABOUT TIME. LET'S REFLECT AND ASK "WHERE DO WE GO FROM HERE?"

As I stated above, during the first few days I spent in Mrs. Waters' Chemistry Block 1 class, it quickly became apparent that time was precious. Throughout my time in her class, I tried to mentally outline all of the types of time-related events that affected teaching and learning—either negatively or positively.

As also noted, from a macroscopic view, I found the actual number of instructional days to be 64 rather than the theoretical 90. From a microscopic view, I was able to see first-hand all of the minute distractions, interruptions, and inefficiencies that steal away valuable time. My research allowed me to construct a taxonomy of these different types of time (table 7.2). It is important to note that, although these time-related terms appear in the literature, there have been few evidence-based studies of them specifically concerning the measurement of transition time, time used waiting for help, time for interruptions—either announced or unannounced, time-on-task, academic learning time, aptitude, perseverance, and pacing. Further, as discussed, the use of the term "engagement" has been problematic both with respect to use and measurement of time. Given today's increased emphasis on teacher evaluations, clarification is needed as to how we measure student learning and classroom observation is requisite to this. The taxonomy provided here is therefore intended to serve as a guide for future research into these relatively unexplored areas.

Further, in my role as a participant-observer in this research, I was able to suggest and implement some changes with Mrs. Waters' help. These were aimed at improving efficiency in real-time. Some of my suggestions still remain unaddressed. A few of these are:

When I was checking out the large, metal cabinets in the storage area, I came across a mobile lab cart (over the years at Beacon, I have developed a reputation for being somewhat of a hoarder). It was in need of some repair but had a sink. A pump and water storage container could be added to it. This would solve a problem in Mrs. Waters' class since running water was lacking. On lab day, many of the experiments required washing out glassware or the use of water. This was an inconvenience and a waste of valuable time. Students had to use the bathrooms down the hall. Further, I thought of mounting a camcorder on the table so that Mrs. Waters could record her experiments and archive them on her website for students who had been absent on one or another lab day—having students who have been absent make up labs is extremely difficult because of the time required for setup, especially when hazardous chemicals are involved. The mobile lab cart was moved to Mrs. Waters' room but remains a project to be completed in the future. There are some interesting computer software products on the horizon that may lead to a future time where some or all science labs are performed in the virtual world.

The fact that Mrs. Waters only had 64 actual instructional days continues to trouble me. A review of her curriculum diary mappings wiki (an outline platform for recording daily classroom instruction and activities) confirmed she had addressed all of the NECAP assessed science standards. However, only the test results, which were to be released in late September, would confirm that adequate time had been spent on preparation. Still, there was material in the course text that could not be covered and arguably would be needed for "college readiness." And, the pacing guide in Mrs. Waters' teacher's edition (Wilbraham, Staley, Matta, & Waterman 2008) was based on 180 instructional days—not 64.

Along the way, I was able to compile a list of "best" practices relative to time and classroom management (see ch. 8: "Mrs. Waters' Best Classroom Practices") that would serve as topics for future professional development sessions at Beacon. Hopefully, I have provided sufficient evidence to support the Doug Lemov *Teach Like a Champion* quote in reference to classroom teaching:

> *". . . Every minute matters"*

Chapter Eight

Mrs. Waters' Best Classroom Practices

> The mediocre teacher tells. The good teacher explains. The superior teacher demonstrates. The great teacher inspires.
>
> —William Arthur Ward[1]

In the film, *The Gods Must be Crazy,* which was released in the US in 1984, an empty Coca-Cola bottle is thrown out of a plane flying over South Africa's Kalahari Desert. The only inhabitants of this completely isolated area are the gentle little people known as the San—or more commonly, the Bushmen. The mysterious object, unbroken by its fall, is found by one of them. He takes it back to his tribe and they wonder why the gods have sent this gift. They have no idea of the purpose of the object, nor do they understand the clear hard material it is made of. One of them blows over the mouth of the bottle, making music. Another finds it a most useful tool for curing snakeskins, and still another uses it as a medicine pounder. Overall, it is found to be a wonderful labor saving device. But the Bushmen have no concept of ownership and unfortunately the gods have sent only one of these gifts. Everyone wants to use it and before long this hitherto very peaceful tribe experiences the unfamiliar emotions of possessiveness, anger, jealousy, hatred, and violence. All of this becomes such a disturbance to their culture that an elder concludes that the gods must have been absent-minded, even crazy, to send down this "evil thing." He then decides to walk to the end of the earth and throw it off. But all question of these negatives aside, when I first saw the film back in the 80's, I found it truly intriguing that a common everyday object in one culture could generate such a level of curiosity in another and thus produce dramatic changes in it.

In my office at school, I have hung many items on the walls from my previous life as an engineer together with various memorabilia from other life experiences. These include awards, golf and racquetball trophies, US Patents for various inventions, a slide rule, musical instruments, college banners, etc. The idea was to cover as much wall space as possible with a wide variety of things that might be as unfamiliar to visiting students as the Bushmen of the Kalahari were with the Coca-Cola bottle—a strategy that was intended to provoke their curiosity and promote opportunities for "teachable moments."

In the urban setting of Beacon Charter High School, many of them have never left the urban confines of Woonsocket, Rhode Island—a place that is not unlike the culturally segregated Kalahari. Fifty percent of them come from conditions of poverty. They have never heard of the sport of racquetball and have no familiarity with something like golf. Most parents can't afford music lessons, although the students are attracted by the few musical instruments I have around the office. Many ask about the four college banners hanging on one wall, each representing a college that I earned a degree from—probably the only college swag they see, is what teachers wear or sometimes display in their classrooms. The slide-rule I have hanging up on one wall might as well be a dinosaur bone. And, of course, the several framed US patents of my inventions from years as an engineer, piques the interest of many of them. But perhaps most importantly, all of this provides the opportunity for me to explain the benefits of working hard in school. It also makes it clear that although math was not an easy subject for me in high school, I wanted to learn and continued to work hard, i.e., if I did it, you can too!

When their attention is attracted by the snapshots from my previous life organized on my office walls, students and other visitors often ask me "Why did you become a math teacher?" I then relate the following story.

During the summer of 2001, while employed as the Director of Engineering for a Rhode Island-based company, I decided I wanted to pursue a master's degree in either engineering or mathematics with the goal of someday teaching nights at the college level as an adjunct professor. In an interview with the chairman of the mathematics department at Providence College, he suggested that I simultaneously take the necessary education courses required for teacher certification by the Rhode Island Department of Education—just in case I wanted to teach at the high school level (education coursework was not a prerequisite for college teaching). I took his advice and subsequently enrolled in both the masters and teacher preparation programs and immediately began doing the necessary coursework. Within just a few months, my decision to move in this new direction would pay off.

In addition to the tragic loss of life caused by the collapse of the North and South towers on the morning of September 11, 2011, this incident dealt a devastating blow to many businesses and industries and my employer was no exception. After twenty-two years in a well established engineering career,

I was laid off. This happened within days of 9/11. For weeks, I scanned the "help wanted" ads for engineering jobs, but with no success. Rhode Island had seen a significant decline in manufacturing for decades so few opportunities existed in my area of expertise. However, in mid-November, an ad in the *Providence Sunday Journal* caught my eye. The Providence, Rhode Island school district was looking for teachers in all content areas. It occurred to me that, since attaining a degree in mechanical engineering required a level of mathematical knowledge well beyond that of a bachelor's degree in math, and since I was officially enrolled in the masters and teacher preparation programs, I might qualify for a high school math position.

A few days after filling out an application with the Human Resources Department of the Providence school district, I received a phone call from them. They had a job for me at a small public high school in the inner city. I would be the only math teacher for about one hundred students for grades 9-12. Classes for grades 9 and 10 met in the Fogarty Building at 111 Fountain St. and for grades 11 and 12, at the Providence Place Mall. Both of these were located in downtown Providence. The school focused on retailing and marketing, so 11th and 12th grade students served internships in various mall stores as part of their high school experience. I was issued a teaching certificate for an "Emergency Teacher of Secondary Mathematics" and instructed to report to the Fogarty Building (see figure 8.1) to begin my first day of work on December 10, 2001.

MY FIRST DAY AS A MATH TEACHER

Monday, December 10, 2001. I arrived at the Fogarty Building at 7:15 a.m. Parking for the school was located in a lot under the building. The poorly lit garage had about 30 spaces available. When I got out of my car, I spotted a security guard stationed by an elevator. He told me that I needed to take the stairs to the third floor as the elevator was broken. The three-story building had been condemned by the city and currently served as a temporary location for the 9th and 10th graders of the Providence Place Academy. Climbing three sets of stairs and exiting the stairwell, I found myself in a fairly large open rectangular space with classrooms located all around the perimeter. In the distance, I spotted a rather tall man nicely dressed in a suit and tie talking to a woman seated behind her desk. My guess was that he was Principal Don Pastine to whom I was supposed to report. No one else was around—neither teachers nor the students—so I knew I had arrived early.

I proceeded to introduce myself and my guess about Principal Pastine was correct. He welcomed me aboard and proceeded to outline to me my schedule. I would be the school's sole math teacher.

Figure 8.1. Fogarty Building, 111 Fountain St.; Providence, Rhode Island.

Today I would teach a 10th grade geometry class, then two 9th grade algebra classes. The fourth block would be my "free" one.

Tomorrow, at the mall site, I would teach an 11th grade algebra II class and the 12th grade pre-calculus class. This would be followed by a "duty" period—standing for 90 minutes supervising students who were working the school's kiosk in the center of the mall. Don told me I would be alternating between both sites, i.e., every other day I would be here at Fogarty. We then proceeded to my classroom.

It was rather narrow, about 15 feet wide by 25 feet long. The far wall was banked with large windows that overlooked Fountain Street three floors below. The room was outfitted with eight five-foot long tables, four on each side and a five-foot aisle down the center. A whiteboard hung on the wall to the left of the only entrance door. Also, there were two empty bookshelves and a standing fan at the back of the room. I noticed right off there were no math books in the classroom and asked Principal Pastine about them. He told me he was not sure where they were but said he was going to leave shortly for the mall site and would check over there. The lack of math books in the room was a little confusing to me since it was now December and school had started in September. Fortunately, I had stuffed my briefcase with algebra and geometry resources and activities for the day just in case.

7:55 a.m. (students arrive). Once Principal Pastine left, I began writing the date "December, 10" on the board along with my name "Mr. Butler" and, "Geometry." This was exciting—my first day as a math teacher! As I was finishing writing on the board, I heard voices in the hallway. Turning my head toward the door, I saw my first student enter the room.

Covered with an amount of gold bling that would be the envy of any Aztec emperor, capped by a flowery do-rag (a kerchief worn over the head) and wearing a matching brightly flowered shirt that fell to just below his knees, Ramiro [pseudonym] entered the classroom. "Hey you fat F---ing bastard! "Where's Mr. Darwin [pseudonym], our sub!"

Immediately, behind Ramiro, three more students appeared, dressed in much the same fashion (Principal Pastine never mentioned school uniforms). These four young men proceeded to the two back tables. Ten more students walked in and took their seats. I began handing out my syllabus. One of the four uniformed students grabbed the standing fan in the back of the room, started breaking the metal spokes off of its protective front guard and handed them to the other three. As I proceeded to the back of the room to retrieve the metal parts, two of the four students quickly threw the spokes upwards like darts sticking them into the ceiling tiles.

Unfortunately, no administrator was around, there was no phone in the classroom, nor was there any other teacher available to watch the class, so I had to deal with the four of them. It occurred to me that this was my very first class and if I didn't deal with these behavioral issues now, I may be sending up a red flag[2]—administration would think I couldn't manage the class. Somehow I make it through the 90-minute session. The next two classes that day went extremely well—there were no issues at all. The fourth block being my "free" one, I made my way to the teachers' lounge where I found an attractive young lady with an engaging smile and soft voice. She introduced herself as the school's art teacher. She explained that this was her second year at the school. I asked her how her day had gone. She explained that she had one very bad class but the other two were great. I asked her if her bad class was the one with the students in uniforms. Her eyes began to tear up as she explained that she had to go out on *stress leave* last year because of that same group of four boys. She also told me that I was the third math teacher since September and that the last one was a nice lady in her late 50's. She had had to be taken to the hospital by ambulance after collapsing from the stress of dealing with the group. Mr. Darwin had been hired as a substitute teacher until the school could find a new certified math teacher. Wow, hadn't I been lucky! Mr. Darwin had developed a relationship with the four boys by talking football for 90 minutes each day and ignoring the fact that it was a math class—and this at the price of ignoring everyone else in the class.

The next day at the mall site, my 11th and 12th grade classes went exceptionally well. My "duty" block required me to supervise the two students who were managing the school's kiosk on the second level of the Providence Place Mall. The school sold various items of clothing and doing this gave them "real world" experience in marketing and retail. The kiosk was located in the open mall space approximately thirty feet from the entrance of a "Victoria's Secret" store. So, not only was I being tortured by members of a gang in an inner city public high school, but also I now had to avoid excessive window shopping for the remainder of the school year.

A few weeks passed and the only problem for me was the geometry class—the "gang of four." They were totally disengaged from school—a pattern that I learned was consistent with these boys in all classes and across all content areas. On Tuesday of Week 3, I was giving a quiz. Ramiro was absent. Twenty minutes into the quiz, the door burst open, nearly hitting my shoulder. Ramiro waltzed in. I ask him to get a late pass from the office: "F--- you!, you F---'ing fat bastard!" He proceeded to place his backpack on his desk. Three girls jump up from their seats and immediately give him a kiss and a hug. Three thoughts enter my mind: (a) I am still waiting for my first pay check; (b) The art teacher was terrorized by this same gang to the point that she went out on stress leave last year; and, (c) Perhaps this career change was not such a great an idea. My reaction…

I calmly asked the young ladies, still attached to Ramiro like Velcro and so suspended in air, to take their seats. Walking calmly past him to his desk and picking up his backpack, I said "Ramiro, watch this." I hear lots of "ooohs" and "ahhhs" from the class. I open the door and start counting "1, 2, . . " while I sway the backpack like one of those big old grandfather clock pendulums back and forth and continue ". . . 3" Then, with a final large thrust of my arm, I release it. It slid down the open hallway like a bowling ball down an alley.

Fifty feet later, strike! It landed between Principal Pastine's legs. He had been standing and talking to the secretary.

In my mind, I had already run through Ramiro's potential reactions to this. I immediately placed both my hands in my pants pockets and allowed myself to be plowed over by him as he ran out the door. Seeing what had happened, Mr. Pastine walked toward him and shouted "Mr. Butler, write him up!" I immediately turned to the remaining three members of the gang and pointed my finger saying "Who wants to be next?" There was dead silence.

Now, the appropriateness of this classroom management technique can be debated, but I never heard a peep out of the three remaining gang members for the next several weeks. Nor did I ever see Ramiro again. Within days of the incident, the faculty elected me Chairman of the SIT (School Improvement Team)—a position I would hold for two years. Further, the three remaining gang members were also removed from the school as a result of incidents that occurred with other teachers.

After this, there were never any issues of student behavior to deal with in the school, neither a fight, nor an argument of any significance.

On that first day, I did learn one valuable lesson, one that would lead me to realize that effective teaching is really very demanding—and complex. It is both a *science* and an *art*. A science in that it requires careful organization, but an art in that it requires creativity. It is by no means enough for a teacher to be knowledgeable of their content area. He or she must also establish an environment that is conducive to effective teaching and learning. All of the students in the class must be actively engaged—even the "Ramiros." The experiences of that first day became the impetus behind my continuous search for "best practices"; specifically, ways of managing a class that promote student engagement in their learning. This is what I referred to earlier as the "aesthetics of learning" (ch. 3).

CLASSROOM MANAGEMENT: MRS. WATER'S BEST PRACTICES

> . . . the most important factor affecting student learning is the teacher. In addition, the results show wide variation in effectiveness among teachers. The immediate and clear implication of this finding is that seemingly more can be done to improve education by improving the effectiveness of teachers than by any other single factor.
>
> —Wright, Horn, and Sanders (1997, 63)

Although recent research has shown that student test scores are now being given greater emphasis in evaluating teacher effectiveness, classroom observations have been, and still remain, the most common and arguably the most heavily weighted method of collecting data on this. The major focus of such observations is on behaviors in the classroom of teacher and students. This is an evaluation area that is often referred to as "classroom management." Historically, it has been the principal that sits in the back of each classroom one or two times a year, observing and making notes. In some cases, an assistant principal or department head will be delegated to do this. In either case, such infrequent snapshots of the classroom by relative strangers all too often contribute to a distorted picture of what constitutes the norm (i.e., participant behaviors can artificially change for a short term).

However, my 79 consecutive days with the students in Mrs. Waters' Chemistry Block 1 class was as far away from such a "snapshot" as could possibly be. They did not perceive me as a stranger in their classroom, as most of them had had me as their Algebra I teacher in ninth grade. And, intrinsic to the relatively small size of the high school (approximately 225 students total) was the fact that our paths crossed on almost a daily basis. My journal notes from Day 25 (March 17, 2011) indicate that:

> Mrs. Waters tells me that Mike (Dean of Faculty/Principal in Residence at the time) had asked her if my presence in the classroom was influencing her teaching. She says she told Mike "not at all." The first few days she was aware of me but has not given it a thought since. She further told Mike that even if she had put on a "dog and pony" show, "no teacher could keep that up this long. After a while, your true colors would show. You can only be a magician for so long before someone sees through the magic."

The duration and frequency of my observations; my acceptance as a native by the students; and, the teacher's assurance that she had not changed her behavior, all speak to the validity of this type of research.

In a work entitled *Synthesis of Research on Classroom Management*, authors Edmund T. Emmer and Carolyn M. Evertson provide the following definition of classroom management:

> Effective classroom management consists of teacher behaviors that produce high levels of student involvement in classroom activities, minimal amounts of student behaviors that interfere with the teacher's or other students' work, and efficient use of instructional time. These criteria have the advantage of being directly observable. (Emmer and Evertson 1981, 342)

Based on this definition, and on the findings in Mrs. Waters' class which were discussed previously, it is clear that she was effective where classroom management was concerned.

In what follows, I discuss classroom *best practices* based on my research as an active participant observer in her class, offer a review of pertinent literature, and comment on my own teaching experience.

The First Few Days

> "you never get a second chance to make a first impression."
>
> —original source unknown

Research suggests that the characteristics and behaviors exhibited to students in the first few days and weeks of the school year by a teacher are predictors of the effectiveness of classroom organization and management throughout the entire school year. The quality of leadership that is brought to bear on problems of student behavior and on instruction, on resolution of student concerns, and on coping with various constraints are all major factors in determining effectiveness. They distinguish effective from ineffective classroom management (Emmer, Evertson, and Anderson 1980; Evertson 2003; Evertson and Emmer 1982; Huitt 1997).

Emmer et al. (1980) state:

> The more effective managers had a workable system of rules and procedures, which they taught their students during the first several weeks. They monitored their students carefully and did not "turn them loose" without careful directions. They did not appear to treat inappropriate behavior differently than less effective managers, but they stopped it sooner. Consequences of appropriate and inappropriate behavior were clearer in their classrooms and were applied more consistently. Thus these teachers established their credibility early and they were predictable. (p. 230).

Among the most important activities of the first week are establishing and teaching rules for appropriate and inappropriate behavior and providing procedures for the routine daily activities (Backes and Ellis 2003; Emmer and Evertson 1981; Huitt 1997). This being the case, the syllabus for the course should be distributed and talked through with the students on the first day of class. At Beacon, every teacher is required to create and distribute a course syllabus. Mrs. Waters distributed hers on the very first day of our chemistry class and we discussed it in great detail.

Although, the literature of peer-reviewed research is lacking in studies that converge on a "best" syllabus format that is linked to student achievement and behavior, there are many online examples of educator syllabi. They share several common elements:

- Course title and applicable semester/school year
- Instructor's name
- Instructor's contact information (e.g., tel., fax, e-mail, website)
- Instructor's brief biography (e.g., degrees earned, certification, teaching experience, professional affiliations)
- Class location (i.e., room number)
- Tutoring/Office Hours
- Required course materials (e.g., textbook, notebook, pencil, calculator)
- Course description
- Standards (e.g., Common Core State Standards)
- Specific learning objectives/skills

- Course topics and order of presentation
- Grading policies (e.g., weighting, due dates, homework)
- Classroom rules/expectations

It is also good practice to attach a form to the syllabus that must be signed by the student's parent or guardian. This means that they have to acknowledge that they have read the document. Keeping parents informed as to what is happening in your class further supports you in your role as a teacher (Pedota 2007).

A CONCISE LIST OF THIRTY "BEST PRACTICES" FOR EFFECTIVE CLASSROOM MANAGEMENT

There is nothing sacred about the number 30. Someone may claim there are 35 or 40 best practices. However, I do claim that the following list provides an excellent checklist for effective classroom management. Further, these can be divided into three general categories, each containing 10 best practices, generally differentiated by a recommended time of implementation. However, it is not my intention here to prioritize an order of implementation within each category:

a) Items to initiate the first few days: #5, 8, 14, 16, 17, 24, 26, 27, 28, and, 29;
b) Items to establish within the first two weeks: #1, 2, 3, 4, 9, 12, 15, 18, 19, and, 30;
c) Remaining items to establish as needed: #6, 7, 10, 11, 13, 20, 21, 22, 23, and 25.

1. *Have a short activity planned for the start of each class* (e.g., problem of the day, vocabulary word of the day). The beginning of class, in addition to the end of a class, are the two times when behavior problems are most likely to occur (Backes and Ellis 2003). Mrs. Waters routinely had a COD (chemical of the day) and/or a state test (i.e., NECAP) aligned POD (problem of the day). Thus students knew each day that they should begin this activity upon entering the classroom. This single strategy meets all three of the effective classroom management criteria defined above (i.e., high levels of student involvement, minimizing interference with teacher's or other students' work, and efficient use of instructional time (Emmer and Evertson 1981). Incorporating a routine activity at the beginning of each class facilitates teacher completion of administrative tasks such as taking attendance and dealing with students that may have been absent the previous class. This strategy can be

employed in all content areas. In math classes, it could be a state-standard aligned math problem; in culinary classes, a "conversion of the day"; in English or Spanish classes, a "vocabulary word of the day"; in visual or theater arts, reading a short biography of an artist or actor and doing a "double-entry journal" activity.

2. *Have the activities and learning objective(s) for the class written on the board.* This strategy provides structure to the class and promotes efficient allocation of time. Students are aware of what is expected in each class. They understand the relevance of what they will learn and are clear on why it is being taught. Mrs. Waters had a consistent daily routine in which she wrote the activities and learning objectives—always in the same general location on the whiteboard. The information written on the board exactly matched her wiki (web-based daily journal of class activities). For the teacher, the benefits of daily journaling include highlighted awareness of the pacing of instruction and confirmation of addressing standards; also this provides an efficient means of fine-tuning instruction for future classes. Below is an example of what was typically written on the board in chemistry class each day:

What will we be doing? 3/24/11
Check pre-lab homework
Do Percent O_2 Lab
"I Am" activity
Objectives: students will be able to
Determine the percent O_2 in Room 205 using experimental methods

3. *A seating plan should be used for at least the first several weeks.* Too often teachers allow students to sit wherever they want in class. Experience has shown that those who are most disengaged from class content typically sit furthest from the teacher. Many of these students are often academically lower performing and/or students most apt to misbehave. In the research I conducted in math classes, I used individual student NWEA Mathematics scores to assign seating. This placed the lowest performing students closest to my primary teaching position and the highest performing ones seated furthest away. The result was statistically significant improvement in student engagement, student performance, and student behavior especially among low performing students and those who had special needs (i.e., had an IEP). Further, I seated the students with IEPs on the ends of rows—a position easily accessible to a resource teacher. Another important reason for assigning seats is that it provides a quick and easy way of learning new names. Backes and Ellis (2003) state "Teachers should commit themselves to learning each student's name during the first two or three days. Calling students by their names make them feel part of the classroom organization" (p. 23). Other reasons

for assigning seating include taking attendance (Pedota 2007), establishing a structure for the class, and sending an early message as to who is in charge.

4. *Provide a physical classroom environment conducive to teaching and learning.* It is important that the teacher have an unobstructed view of students at all times and vice versa (Pedota 2007). And, ideally, the tables and/or desks should be arranged such that the teacher has sufficient access to every student. It is also important that students be seated in a position that allows them a proper view of the board (in classes where one is used). As noted, Mrs. Waters had her chemistry class seated in a U-shaped pattern. This allowed her an unobstructed view of students and vice-versa. It also seemed to help promote student engagement. Temperature and lighting also affect student learning. As a student in her chemistry class on the first day, I sat in a seat that was placed at an angle to the board such that the glare of the sunlight from a window obscured the writing on the board. I never sat in that seat again, though other students did. In the course of the semester, it was common for students to have to get up and move closer to the board in order to write down notes. For this and other reasons, I highly recommend that teachers occasionally walk around to check the notes that students are taking down. Some students have vision problems and are embarrassed (or sometimes too lazy) to ask for a seating change.

5. *Give a pre-test at the beginning of the course to establish baselines for knowledge of course content.* As concerns, both RTI (Response to Intervention) and the new Rhode Island Model Educator Evaluation System, establishing a baseline for student performance is important. It allows one to clearly monitor their progress. A pre-test serves this purpose and should be administered as early in the course as possibly—preferably before any significant instruction has occurred.

6. *Frequent monitoring of student progress.* Where RTI (Response to Intervention) is concerned, it is important to keep track of student progress both academically and behaviorally. Students who fall behind need to be provided with the requisite support as early as possible (Huitt 1997).

7. *Assign group members/partners and model roles.* In those chemistry classes of Mrs. Waters that I observed and participated in, she always assigned group memberships for group activities. Further, she varied these each time a group activity was used—we never knew who would be partnered with whom. One of her strategies was obviously to assign at least one academically strong student to a group. Further, members of the groups had roles (e.g., scribe, time keeper). It is worth noting that over the 79 days I spent in the chemistry class, no student ever complained openly about their group assignment even though they had no input into their placement in a given group. This strategy sets a clear tone in the class. It makes it clear that it is structured

and helps reaffirm the authority of the person in charge. It also promotes team building. There may be some good reasons why students did not complain or question their group placement. I would suggest, first, Mrs. Waters assigned group membership with such consistency that it became an expectation on the part of the students; and secondly, she had developed an environment of respect and had good rapport with them.

8. *Sign-out/sign-in and one student out at a time*. It is important to minimize the amount of time students are out of the classroom and the number of them who exit the class at the same time. This detracts from precious ALT (Academic Learning Time). A sign-out/sign-in sheet should be utilized to monitor the frequency and duration of their absences, and record their reasons for leaving the class (Pedota 2007). Teachers should look for patterns of behavior and follow RTI protocols as appropriate.

9. *Provide multiple activities with minimal transition time.* Multiple activities will help keep students engaged (Pedota 2007). They can assist with transitions such as passing out papers. For some, 80-minute classes (or longer) are too long. They cannot stay focused on the same activity for such a length of time. This may be one reason why many of them look for an excuse to leave the classroom.

10. *Assign at least a few "flexible" homeworks and projects*. One major finding in my research in Mrs. Waters' chemistry class was that there was a significant improvement in student engagement and in quality of submitted work when students were given a voice in designing their assignments. For example, when they were given an assignment to create their own homework related to a chemistry topic, one student baked cupcakes frosted with the names of the elements from the periodic table and arranged them on a tray to form compounds—a four-hour homework assignment by her choice. Another one was studying the reaction of yeast in the baking of a pizza and photographed each step of the process. She made a poster, and wrote a short paper—another four-hour exploration, but by her own choice. Near the end of the semester, on another flexible assignment, one student spent 20 hours producing a painting on the life cycle of a star. Even lower performing students were submitting portfolio-quality work on simple homework assignments when they were given this flexibility.

11. *Create or use study guides aligned with PowerPoint or film presentations.* It has always been my experience that simply having students view PowerPoint presentations or films without including an engaging hands-on activity causes some of them to be become disengaged. Thus, I create and use study guides that require the student to watch and listen carefully so as to be able to answer relevant questions—this requires a higher degree of DOK (Depth of Knowledge) than simply viewing information. In an exit survey

given to all 11th graders who had taken chemistry, 68 percent indicated that they preferred completing a study guide to just copying notes from the PowerPoint slides; 76 percent felt they learned more by completing the study guide than just copying down notes. When I first shared my philosophy with Mrs. Waters on this question of incorporating study guides, I also presented her with two examples using the excellent PowerPoints she had posted on her website. She immediately used them for her class and created study guides for all of her other PowerPoints as well. This is a strategy that has now become routine for her.

12. *Make course materials available on the teacher's website (e.g., Digication) whenever possible.* Teachers should utilize their website for posting resource materials such as PowerPoints, handouts, project descriptions, sample student work, etc. This is especially useful for those students who may have been absent from class.

13. *Use unison responses.* As a means of attaining high levels of student involvement, increasing instructional time, and providing an efficient means of formative assessment, have all students in the class occasionally respond to questions or problems by simultaneously holding up their answers on a piece a paper or on a mini-whiteboard. They can also use hand signals or sign language, or they can use texting devices with associated software (i.e., student response systems). This makes it possible to quickly assess which students are "getting it," and which are not. Such response methods provide significantly more opportunities for everyone to respond than if the teacher simply addresses one at a time. Further, students are less likely to misbehave if they are actively engaged in this whole-class activity.

14. *Promote use of the students' Agenda/Planner.* The student agenda or some kind of student planner is a valuable tool for teaching organization skills. Teachers should routinely check that students are writing down homework assignments and project due dates.

15. *Promote good note taking and notebook organization.* Lower performing students often say they don't understand something, but when I ask to see their class notes, it is clear that they did not take any, or that they are incomplete and/or indiscernible. Both my experience and existing research suggest that higher performing students are generally more organized where this is concerned than are others. Their notes are neater and more complete. I require students to maintain a notebook and I make it a personal goal each year to work with them on developing these skills. Mrs. Waters had clear expectations for notebooks outlined in her course syllabus she handed out the first day of class. She checked our note taking and 3-ring binder chemistry notebook (or three separate notebooks) on a number of occasions during the semester. And, she had a specific requirement for labeling and organizing

the binder sections. Mrs. Waters and I agreed that she should put together a model notebook consisting of an entire semester of work that could be passed around on the first day of the course to demonstrate her expectations for a neat, complete, and correct product.

16. *Don't allow backpacks on the desks.* This is in line with the need to ensure that the teacher has an unobstructed view of students. Backpacks provide a nice barrier for texting on their cell phones or working on things other than what they are supposed to be doing.

17. *Don't allow students to get up early at the end of class and stand by the door.* Students standing by the door at the end of class feeds the perception that there is not "bell-to-bell" teaching going on in the classroom, students are disengaged and can't wait to leave, ALT (Academic Learning Time) is being wasted here too, and students do not respect the teacher as the one who dismisses the class (Pedota 2007). Both the beginning and the end are likely times for behavioral problems—students are off task (Backes and Ellis 2003). If class occasionally finishes early, the teacher should have tasks pre-planned that fill the void such as an "exit ticket" activity. Make every minute count!

18. *Have a sub folder ready in case of teacher absence.* Like everyone else, teachers get sick. They have emergencies, or may have to be absent from school for other reasons. Consequently, coverage must be provided. It is therefore highly recommended that a "sub folder" of course related activities be readily available. It can ensure some degree of continuity or ensure that at least a modicum of learning takes place in the teacher's absence; also, it can help minimize opportunities for inappropriate behavior. Mrs. Waters always had a game plan when she was going to be out. In the emergency situations I have described, she took responsibility for calling an administrator or myself for help in running off copies from her sub folders. No class was ever left adrift, in her absence no sessions were merely run as a study hall. There was always at least some type of learning relative to chemistry going on.

19. *Keep the grade book current.* It is important for all stakeholders (i.e., teacher, student, parent/guardian, special educator) to be aware of the current status of the student's academic performance. Interventions can then be made as early as possible if that student starts to fall behind. Further, this helps prevent situations in which parents are caught by surprise when they discover that their child is not performing as well as they may have been being told. Hence, the need to keep the grade book current. I found Mrs. Waters to be exemplary in this. Several students in her Chemistry Block 1 class checked their grades online on almost a daily basis. This is further evidence that they were engaged in their learning and desired to achieve.

20. *Keep parents informed.* Do not ignore or put off attempts by parents to communicate with you. Make yourself available at the earliest possibility.

In most cases, involving them will help support you in your role as a teacher (Pedota 2007).

21. *Use surveys and reflections to get student feedback on feelings toward assignments and to assess one's teaching from the students' perspective.* By using surveys in her chemistry class, Mrs. Waters learned that out of the twelve different activities she routinely used during the semester, lab experiments were their favorite. This was followed by the Jeopardy game, flexible homework assignments, and extra credit opportunities. Students also reported that taking PowerPoint notes without a study guide and writing a formal lab report were their two least favorite activities. With this valuable data in hand, Mrs. Waters can adjust her pedagogy accordingly. In my own classes, I have required math students to attach to their assignments a paragraph of three to five sentences which reflects their feelings about both their performance and my teaching: How did they feel about the assignment? Did I provide sufficient instruction? Did I move too fast? Using this strategy over the years has given me valuable insight to my effectiveness as a teacher; it has also drawn my attention to specific assignments that were in need of improvement.

22. *Provide extra credit opportunities.* My research has shown that a significant number of students welcome the opportunity to raise their grades by doing extra credit assignments. Many students will do more if given the opportunity. If an assignment is meaningful and relevant, it is yet another means of increasing student engagement and deepening learning. Mrs. Waters consistently provided opportunities for extra credit assignments throughout the semester and many students responded.

23. *Use NWEA data or other data sources to differentiate assignments.* In preparing for the NECAP Science assessments, Mrs. Waters drew on student NWEA science performance data to differentiate assignments. She was able, for example, to assign specific readings to topics in their weakest areas. This was also done for the summer reading assignment for the upcoming eleventh 11th graders.

24. *Don't allow heads down on the desk.* I propose that there are three reasons why a student will put his or her head down on a desk: Not feeling well, being tired, or being bored. To solve these problems, in the first case, there's the nurse; in the second, if the problem persists, a phone call should be made to the home, in the third, question your pedagogy!

25. *In planning activities for academic classes, consider the aesthetics of learning.* As a means of increasing student engagement, use some activities that appeal to their aesthetic interests (e.g., using their skills and passion for art in content-related assignments).

26. *Teachers should model and preserve the dignity of students at all times.* "Even during a reprimand, teachers should preserve the dignity of the student.

When the reprimand is over, teachers should not hold a grudge and punish the student further by ignoring him/her for days afterwards or by making nonverbal gestures. This causes problems to grow and causes learning not to take place" (Backes and Ellis 2003, 23). Even during student presentations that fall below expectations, find at least one good thing to say.

27. *Teachers must always model appropriate behavior and use appropriate language.* "Teachers should model appropriate behaviors such as patience, a sense of humor, courtesy, and manner" (Backes and Ellis 2003, 23). Appropriate behavior includes appropriate dress/appearance. Further, it is important for teacher credibility as role models to refrain from ethnic jokes or remarks, sexual innuendos, etc. and also to refrain from profanity. Also, Backes and Ellis (2003) state "Certain physical or verbal responses made by an instructor seem to attract inappropriate behavior by students. These responses may include threats of consequences that can't be administered and extreme physical reactions such as turning red in the face, yelling or otherwise overreacting to relatively minor offenses. Teachers should avoid these overreactions, so students don't get the satisfaction of seeing them lose their cool" (p. 24). Further, as concerns instruction, teachers must avoid telegraphing negative feelings toward the content being taught. "Effective instruction communicates a sense of purpose as well as a sense of responsibility for the value of what one teaches" (Bowman 1983, 117). "When caught in the all too familiar crunch from teaching from prescribed curriculum materials which are perceived to be only of secondary importance, teachers often unwittingly telegraph deep-seated feelings of frustration, embarrassment, and indignation. Predictably, these signals trigger telltale disruptions in the learning environment" (p. 117).

28. *Young teachers need to be careful to keep their personal lives personal.* "Doing otherwise confuses high school students and makes them feel that the teacher is just another friend, companion or confidant. This close relationship can be dangerous and cause employment and sexual harassment issues" (Backes and Ellis 2003, 25).

29. *Teachers should use electronic communication (e.g., emails, texting, Facebook, other social media) with students for academic purposes only.* It is important to maintain an appropriate teacher-student relationship. Inappropriate communication can be dangerous and lead to employment and sexual harassment issues. However, the proper use of social media can greatly enhance student engagement in their learning.

30. *The quality of teacher-student relationships is the keystone for all other aspects of classroom management (Marzano and Marzano 2003, 6).* "Teachers with effective classroom management skills are aware of high-needs students and have a repertoire of specific techniques for meeting some of their needs" (p. 11).

HOW CAN ADMINISTRATION HELP WITH CLASSROOM MANAGEMENT?

- A wise man from another planet once stated "The needs of the many outweigh the needs of the few" (Spock in *Star Trek II: The Wrath of Khan*). When a student or students are disruptive in the class to the point that they are stealing valuable ALT (Academic Learning Time) from others, and all appropriate measures have been tried by the teacher to address the behavior problem, it may be necessary for the teacher to remove the student(s) from the class. By no means, should this be taken by the administration as an indication of a lack of control on the part of the teacher over the classroom unless a pattern develops. Sometimes, *not* removing the troublesome students is poor classroom management.
- Minimize the number of announcements that are made over the PA system during class periods.
- Minimize the number of phone calls to the classroom. They often require the teacher to stop in the midst of instruction and address some issue that could have been handled before school begins, between classes, during a teacher's free period, lunch period, or after school.
- Inform teachers well in advance of field trips, college visits, etc. and have students sign-up for such activities in advance. The teacher needs to be aware of who will be present or not in the class.

CONCLUDING REMARKS: THE LAW OF CONSERVATION OF ENERGY

There is a fundamental principle in physics known as the Law of Conservation of Energy in a System. It states:

> *Energy can neither be created nor destroyed: it can only be transformed from one state to another.*

A parallel exists here in respect to classroom management. The more energy that is put into establishing rules and procedures in the first few weeks, the less will be required for classroom management during the rest of the school year.

As concerns, a RTI (Response to Intervention) three-tiered model, the more energy put into Tier 1, the less will be required for Tier 2 and Tier 3 support.

NOTES

1. William Arthur Ward (1921–1994), an American author, is one of the most frequently quoted writers from his books (e.g., *Fountains of Faith*), and his more than 100 articles, poems and meditations. This particular quote from Ward was taken from Gerald Anderson's *Achieving Teaching Excellence: A Step-by-Step Guide* (2009, 3).
2. The term "red flag" has several meanings. In this context, it refers to an attention getting signal such as a warning.

Chapter Nine

Teacher of the Year—It's the Climb

MIXED EMOTIONS

In chapter 6, "The State Science Exam," I discussed the mental state of students and the issue of learned helplessness. But teachers can also fall into the learned helplessness trap:

During my 79 days with Mrs. Waters, it was rare to see signs of stress or frustration in her. Quite the contrary—I rarely found her to be without a smile. However, I do recall one day when she questioned whether all of her effort was worth it. Given her busy schedule, her balancing the three roles of teacher, mother, and wife, I was surprised that there weren't more "down" moments or days.

She was working many hours correcting papers, keeping her grade book and curriculum mapping diary wiki current and preparing for the next day of classes. Her kids had been sick for a few days and her husband Chris was suggesting that she get a masters degree to become more marketable so as to earn a higher salary. Chris is a highly successful CAO (Chief Accounting Officer) who does business on an international level. As with most private sector jobs, he is paid at a level which is commensurable with his skills and responsibilities. This stands in relatively sharp contrast to public school teachers who are generally paid on the basis of years of service.

There are over three million public school, K-12 teachers in the United States and approximately 80 percent of them are women. And, a significant number of these teachers have been in the same position as Mrs. Waters—balancing the roles of teacher, mother, and wife. Yet one rarely hears or reads about the many highly effective female teachers across the country, these women who successfully manage both career and home. On the contrary, a disproportionate amount of media attention is given to negative events in the field of education such as inappropriate teacher-student relationships.

So, as Mrs. Waters continued to question the value of her effort, I told her that whenever I had similar "down" moments I would ask myself: Why do I teach? Because I want to positively affect students' lives or is it because it's the job that provides health benefits?

I reminded her of the flexible homework she had assigned and the great student response to it—Sarah's four hours making periodic table cupcakes, Brittany's several hours studying yeast as she baked pizza, and Robert's twenty hours painting "The Life Cycle of a Star." These students spent hours on a simple homework assignment because a highly effective teacher awakened in them an interest in learning. I further reminded Mrs. Waters of the surprising number of girls in her science classes who were now expressing a desire to pursue STEM (Science, Technology, Engineering, Mathematics) related careers even though they had originally come to Beacon for art. I believe that the excellence of Mrs. Waters' engaging teaching played a significant role in their decisions. And, this was quite a welcome outcome considering the national call to increase the number of women in the STEM workforce.

I further explained to her that it has been my experience, not always, but more often than not, that effort and perseverance do eventually reap rewards. This is the reason why I keep visiting the Foxwoods Casino—have to get lucky one of these days!

I ended our conversation that day with the following quote: "The highest reward for a person's toil is not what they get for it, but what they become by it"—a quote attributed to John Ruskin (19th century writer, art critic, social thinker).

September 6 (Tuesday), 2011. At today's faculty meeting, Principal Mike Skeldon announced that Beacon was going to nominate a teacher for the "2012 National Teacher of the Year." Mike, along with Beacon's "Cabinet" (comprised of the school's administrators) had established criteria for the nomination of candidates. Based on these, they had prepared ballots for a full faculty vote. The faculty was instructed to submit their completed ballot before leaving campus today.

One day later. September 7, 2011 was a special day for Mrs. Waters—Mike Skeldon informed her that she had been nominated by Beacon's faculty for Rhode Island's "Teacher of the Year." She was now one of several nominees from amongst the approximately forty-eight Rhode Island school districts who would be considered for nomination by the state for this award.

COMMUNICATING WITH THE YOUNGER GENERATION

Texting with Mrs. Waters

I've owned a cell phone for over ten years but had never sent a text message with it until I met Mrs. Waters. She must spend a lot of time texting because messages from her appear at lightning speed, one right after the other. (I think the younger teachers practice on this during faculty meetings). A lot of the text messages I receive often contain strange combinations of letters and abbreviated words like "BTW," "LOL," and "TTYL." I was therefore forced to purchase Randall C. Manning's *Texting Dictionary of Acronyms*.

Responding to text messages is not easy for me because of my fat fingers! They tend to hit the wrong letters and my phone automatically corrects the spelling—suddenly some new word appears that I never intended to key in! Then there is the problem of frequently hitting the *Send* button prematurely. Not only am I slow to respond, but sometimes I don't even understand my own text messages! Here is a message that came in from Mrs. Waters one day when we were texting:

> Ps you're a slow typer
> In the time it took for you to reply I sent two emails
> updated my status on Facebook
> and commented on my friends pic

So, given her critique, I have chosen to list her text messages to me here below without including my responses. If I don't understand my own responses, how can I expect the reader to? This decision is a result of that mental state called learned helplessness (ch. 6, "The State Science Exam").

THE CLIMB BEGINS

September 7, 2011 (Wednesday), 8:16 p.m. I receive the following text message from Mrs. Waters:

> Hey so I won today Teacher of the Year nominee
> I'm honored that my colleagues picked me
> The paperwork is 17 pages long due
> next Friday There is no way I can do it
> without your help it's overwhelming

As part of the application process, in addition to listing education and teaching experience, each candidate must write essays in response to a set of questions relative to their professional experience and philosophies. The application deadline is Friday, September 16, just nine days from now. Over the next several days, Mrs. Waters completes it. I was fortunate, along with Principal Mike Skeldon, to have a preview of her responses to the questions. Mike found a couple of grammatical errors and passed them on to her for correction. The application package was successfully delivered to the Department of Education on Friday—ahead of the deadline.

October 12, 2011 (Wednesday). Mrs. Waters receives a phone call from RIDE (Rhode Island Department of Education) informing her that she is one of six finalists for the Rhode Island 2012 Teacher of the Year. She has been invited to RIDE for an in-person interview at 5:00 p.m. on Wednesday, October 19. She calls me with the good news. Having read her application, I was not surprised that she would be considered a finalist.

October 18, 2011 (Tuesday), 12:45 p.m. (teachers' lunchroom). It is the day before her interview. We are having lunch with several other faculty members. Mrs. Waters, now seven months pregnant (and showing!) expresses a concern about her appearance and asks what she should wear to the interview. Dana and Robin, two other faculty members, who are in attendance and both also pregnant, offer their opinions.

At one point, Dana and Mrs. Waters simultaneously start calling everyone's attention to the small movements in their respective stomachs as the unborn are kicking. Not to be left out of the conversation, I point out that I feel some kicking in my stomach too but it's from the buffalo chicken sandwich I just consumed! A discussion ensues. It centers on the question of whether or not Mrs. Waters' pregnancy would be a factor in the RIDE search committees' decision-making process.

5:45 p.m. It's the night before her presentation at RIDE. I receive a series of text messages from her:

> I am so nervous about tomorrow…can't think about anything else…
> I actually served box mac and cheese to my kids tonight something I never ever do lol
> How many people do you think will be at the interview?
> I am making myself sick…what if I say the wrong thing…what if the others r so much better
>
> I'm trying to get Madi to sleep lol
> I'm reading the application to her lol

October 19, 2011 (Wednesday), 4:47 p.m. (the big day). It is thirteen minutes before Mrs. Waters' scheduled interview. It's been pouring all day. I receive the following text messages.

Got here with ample time
But I'm soaked lol
But I have black on so no one will no I'm wet lol
So nervous
Should I go up early or wait?
K going up I feel like I'm going to vomit
Thanks hope I don't let anyone down

5:50 p.m. Mrs. Waters calls me right after the interview and briefly describes the event. Seven people were seated at a table. A sheet containing four questions was taped to the head of the table. The format of the questions was exactly the same as the one that Beacon uses when interviewing teachers for open positions. She first experienced this format for questioning when she interviewed for the science teacher position at Beacon, and has since sat on the other side of the table as an interviewer. She was therefore very familiar with the process. The questions were evidently easy to respond to and she was confident she had answered them fully. The interview was capped at the fifteen-minute limit. She was nervous going in to it, but felt good coming out. She was told that six finalists were being interviewed and that the search committee would stay that evening until they had chosen the three who would be invited back to give a presentation next week. Within the next two days, she would receive their decision.

8:25 p.m. (just hours later). I receive the following text messages:

Department of Ed just called I missed the call
the lady asked if I could call her back
I did but went right to vm
Is this good or bad?

Just called me and said I did fantastic and she is calling Mike
Next Tuesday 10 minute presentation at 4:30
Ahhhhh so excited
She is sending me a question in an email
This is crazy!!
Looks like another sleepless night lol

So, Mrs. Waters is now one of three finalists for the Rhode Island 2012 Teacher of the Year. Not bad, considering there are over 14,000 public school K-12 teachers in the state!

For selfish reasons, as an educational researcher, I am feeling pretty good, too. It has now been four months since I completed her Chemistry Block 1 class. During my 79 days in her class, I had compiled an impressive list of "best teaching practices." It was supported by evidence in the form of student work and feedback—data that all support the conclusion that Mrs. Waters is a highly effective teacher. Given that she was one of three finalists, my positive perspective on her classroom practices and educational philosophies had now been validated by an external source, a team from the Department of Education. Reflecting back on this, I now realize that I really didn't know Mrs. Waters so well prior to undertaking my research. She just seemed like the obvious choice of a science teacher to work with on my research. But what were the odds that she would end up being one of three finalists for Rhode Island's Teacher of the Year when there was a pool of over 14,000 teachers? I'm feeling very lucky—looks like a trip to Foxwoods tonight!

Now, Mrs. Waters would have to prepare a presentation. However, she can't begin until the question to be addressed is received from the Department of Education. She keeps checking her email. Every hour that passes seems like an eternity. This is a source of stress because she would like as much time as possible to prepare a good presentation. Thursday also passes with no communication from the DOE. The weekend is approaching and it would be nice if Mrs. Waters could at least have the weekend to work on her presentation.

October 21, 2011 (Friday), 1:19:09 PM EDT. It is Friday. Finally, shortly after lunch, the question arrives via an email (figure 9.1).

From the email heading (not shown here to maintain confidentiality), Mrs. Waters determines that her two competitors are from two large school districts, Coventry and Warwick. I recognize the name of the teacher from Coventry High School. I had taught math there a few years ago, and Julie Boyle, a teacher in the English department, was highly regarded by the faculty—so, I am not surprised to hear she is a finalist.

Choosing a Topic

Mrs. Waters asks my opinion. I suggested several that she might discuss—topics that I knew she was very familiar with such as RTI (Response to Intervention); the new Rhode Island Model Educator Evaluation System that she had volunteered to pilot; the results of flexible homework assignments; and, her own personal story of being a high school dropout. Mrs. Waters decides to discuss the national high school dropout rate problem and her strategies for promoting student engagement. She prepares a 16-slide PowerPoint presentation and feels confident she can deliver it within the 10-minute limit. Her presentation is entitled "In Search of a Field of Dreams[1]: Engaging Students."

October 25, 2011 (Tuesday), 4:30 p.m. (presentation day for the three finalists). Mrs. Waters arrives at the same room she had previously been to along with her competition. The audience has now grown to eight in number. The three presenters pick numbers out of a hat in order to determine the order

[Heading deleted for confidentiality]

Subject: Interview Question October 25, 2011

Dear Teacher of the Year Finalists,

First, congratulations! We are thrilled to have you as our top three finalists representing Rhode Island educators.

On Tuesday, October 25th from 4.30 -6.00pm we will conduct final interviews to determine our Rhode Island Teacher of the Year. The interview process will be conducted as follows:

I. Welcome and Overview of the Interview Protocol
II. Individual Presentations
III. Panel Interview
IV. Next Steps

The presentation question will be answered by each finalist separately. Each finalist will pick a number to determine the order in which she is to present (1st, 2nd or 3rd). Each candidate will have 15 minutes for their presentation (five minutes will be allocated to set up for the presentation and ten minutes to present their topic to the panel).
The presentation topic is specifically broad. You may bring in materials or visuals to enhance your presentation. There will be a projector and laptop at your disposal.

Presentation:
"Select a topic of importance on an issue of both personal/ professional importance within the educational field."

The second part of the interview will be a "Mock Press Conference Interview." Each candidate will be given an opportunity to address three (3) pre-selected questions relating to state and national education policy. Each finalist will pick a number to determine the primary question they will be asked to answer (question 1, 2 or 3). Candidates will have 3 minutes to answer their primary question and 1 minute to address their colleagues questions.

If you have any questions, please contact me.

[Contact info deleted for confidentiality]

Figure 9.1. The Interview question arrives.

of presentation. Mrs. Waters will go first, Coventry second, and Warwick last. However, each presentation is done separately (i.e., the candidates do not get to view the other presenters) so Mrs. Waters has no idea what the others will have chosen for topics and no sense of how their presentations went. After Warwick presented, the three got together for the mock press conference. They picked a random question and were given a strict three minute time period in which to respond; then the other two presenters each had one minute to comment on the question.

6:16 p.m. (my phone rings). Mrs. Waters calls me following her presentation. She says that things didn't go as well as she would have liked. She thought she did well with her PowerPoint presentation, but when she sensed she was running out of time, she could only briefly touch on a few remaining slides. It had been impossible to cover all of them. However, she was frustrated with the question she had drawn during the mock press interview. It was long and detailed—long enough that it had to be read off a paper by the mock reporter. It had to do with NCLB (No Child Left Behind) and what Mrs. Waters would say to Senator Jack Reed (Democratic US Senator from Rhode Island) in regards to it. She said she stumbled her way through it. By contrast to this, however, she thought she had much more knowledge on the other two questions—the ones posed to her competition. However, she felt that she did a much better job on the one-minute comments to the other two questions.

The question in her mind was how much weight would be given to each of the components: the application essays; the in-person interview of the six finalists; and now, this afternoon's presentation and performance on the mock interview? Mrs. Waters tells me that before leaving the room, all three contestants were instructed to review their application for grammatical errors only—they were not allowed to change the content and had to submit any changes by Thursday at noon. They were further informed that the search committee would stay until they decided the Rhode Island Teacher of the Year so as to be able to submit the name of the winner to Washington for the National competition. However, the name would be kept secret until the presentation by Rhode Island Governor Lincoln D. Chafee and Commissioner of Education Deborah A. Gist. The event would take place sometime in the next few months. Only the principal and a member of the winner's family would be informed the night before the presentation. The Governor and Commissioner, together with the media, would show up at school and surprise the Teacher of the Year.

. . . so the sleepless nights, anxiousness, and mind distracting moments will continue.

November 2, 2011 (Wednesday). Still no word on the winner, but all Teacher of the Year nominees, along with their school district superintendents, have been invited to the RIPEC (Rhode Island Expenditure Council) 68th Annual Dinner being held at the Rhode Island Convention Center in downtown Providence. Arne Duncan, US Secretary of Education is the keynote speaker. Several hundred attendees are present, including Rhode Island's Governor Chafee, Commissioner of Education Gist and other dignitaries. The location of the dinner had not been announced until a few days ago because protests had been planned on the occasion of Arne Duncan's visit. Security was high. Again, only superintendents were invited from the school districts, so Dr. Pilkington, our former principal (Beacon School District does not have a Board approved superintendent) unfortunately could not be there with Mrs.

Waters because of a prior commitment. A request was submitted for Principal Mike Skeldon to attend in his place. No response to the request had come in, so Mike showed up with Mrs. Waters anyway.

They made their way past the two dozen or more protestors who were shouting "Hey hey, ho, ho, Arne Duncan has to go" and who were carrying signs that read "Where Arne goes, schools close," and "Arne Duncan, Dr. Death." Just inside the building, Mike was stopped from entering because he was not on the invited guest list, nor was the name of former Principal Pilkington. Fortunately, after twenty-five minutes of intense negotiating, a photo ID check, a retina scan, some blood work, an MRI, and a urine test, Mike was allowed to enter and given a seat.

The search committee for Teacher of the Year was also invited as were several past winners. Mrs. Waters told me she spent much of the evening trying to get a clue as to who the winner was. But everyone was wearing their poker faces. There was some discussion by past winners who explained that the Teacher of the Year entails a lot of work, several days away from the classroom. Mrs. Waters told me that she didn't think Mike would be too happy to hear that. She tells me that Arne gave a good speech. The dinner was tasty "Just glad I was invited, Mike got in, and we didn't have to pay the $300 per plate ticket to the event." The event ended with no further indication as to who had won.

. . . so the sleepless nights, anxiousness, and mind distracting moments will continue.

November 7, 2011 (Monday), 7:30 p.m. I receive the following text message from Mrs. Waters:

> One teacher 2011 was surprised on Dec 17
> and 2010 was December 3

Mrs. Waters' is due to deliver her baby on December 8 but her doctor doesn't want her to go more than one week beyond the due date, so December 1st is looking like the special date. The timeline is worrying her. The climb continues. We are now nearing the summit. Breathing is becoming increasingly difficult.

While We're Waiting: Some Waters' Philosophy

I asked Mrs. Waters' if I could include the following excerpts from the application she had submitted to the Rhode Island Department of Education. I believe they provide valuable insight to the thinking of a highly effective teacher. The following are her responses to all six questions submitted (Sections III through Section VIII of the Rhode Island Application).

What were the factors that influenced you to become a teacher? Describe what you consider to be your greatest contributions and accomplishments in education (from Section III of the application: Professional Biography)—I grew up with my parents who had drug addiction problems. I remember the day the police came to our home and arrested them. My little sister ran into the woods because she was so scared. The police just left my sister and me at home. I was in 10th grade at the time and decided to quit school. My grades were excellent but my home situation was such that I knew I had to move out. I got my mother to go to the school with me to sign me out. We met with the guidance counselor. He asked me why we were here. I told him I wanted to sign out. He opened his desk drawer, and without saying one word, he thrust the necessary forms in front of us. My mother, with her mind in another place, signed the papers. The uncaring nature of the guidance counselor—the speed with which he wrote me off as a student—these were life-changing. I recall thinking that if I were that counselor or a teacher, I would have tried everything in my power to work with that student.

I subsequently moved out and took on two jobs, working from 7:00 a.m. to 11:00 p.m. every day to pay for the rent on an apartment and manage the essentials of life. I was able to squeeze out enough time to earn my GED and immediately began classes at CCRI. I had aspirations of being a doctor someday, but decided I wanted to be a caring teacher who could make a difference in somebody's life. Today, I teach science in an urban charter high school. I am fulfilling my dream as we have more than our share of students from troubled backgrounds—some of them mirrors of my past.

I would like to tell you about my "greatest" contributions and accomplishments in education, but I am hoping they're ahead of me not behind. Let me instead discuss what I take to be some of my major accomplishments to date:

Beacon Charter High School has a focus in the arts. Students choose a culinary, visual, or theater arts discipline for their four-year high school experience. Science is not the reason why students come to our school. So, I am especially proud when students exhibit enthusiasm and are clearly engaged in the science content area—especially the girls and my special education students. I have even had a few students express an interest in science related careers such as the medical profession. I am proud of the continuous improvement in my students' performance on the NWEA (Northwest Evaluation Association) science assessments that are given to each Beacon student twice a year. I am proud of all of the effort that my students put forth in preparing for the state's NECAP science assessments. I feel we are on the verge of setting achievement records on the NECAP, if not this year, then next— it seems inevitable! And, each semester, I feel my teaching is getting better as is my ability to inspire students of all backgrounds and learning abilities. I am humbled by the support lent to me by my administration and colleagues and by their choosing me as their nominee for this prestigious award.

Last year, I volunteered to have an educational researcher observe one of my chemistry classes all semester. He was doing an ethnographic study of student engagement and classroom management. My students and I were observed for 79 consecutive days. He recently informed that he is publishing his findings in a book entitled *Who's the New Kid in Chemistry?: Exploring Uncharted Waters*. I understand it's a very positive perspective; and, again I am humbled to think that I may positively affect readers.

I am also proud of my work as a volunteer for the state's pilot of the new educator evaluation system and on the school's RTI (Response-to-Intervention) Handbook committee.

Lastly, I am proud of my ability to balance career and family. I am a happily married mother of two. I believe that I have been able to successfully multi-task between career and family obligations.

Describe your commitment to your community through service-oriented activities such as volunteer work, civic responsibilities and other group activities (from Section IV of the application: Community Involvement)—Today, community service is an important component of my life. As discussed in Section III, I moved out of my family home at sixteen. Living alone, I thought things were very bad for me. I remember that first Thanksgiving alone. My aunt called to check up on me and said "Jess, I want you to come volunteer with me to serve Thanksgiving dinner at the homeless shelter in Providence." I was mad at her for asking me to go and help other people when she should have been helping me, but reluctantly went and spent the day serving dinners. It was truly an emotional experience, one that I have never forgotten—another life changing event—a real "eye-opener." I realized that no matter how hard times may seem to me, they pale in comparison to the hardship of the single adults and families with children I served that day.

In my first year at Beacon, I initiated, and have thereafter continued, our annual Thanksgiving food drive. Many of the students I teach are in a place similar to that in which I found myself at age 16, so I encourage them to reach out to others and help them. This annual food drive collects over 1500 non-perishable food items. They are used to make Thanksgiving baskets for families in the Woonsocket area. Teachers donate gift cards for the baskets and on the Thursday before Thanksgiving we pack it all up and take it to Family Resources, a local social service agency, where they are dispensed to families in need. We do a similar drive for food baskets at Christmas, too.

Another issue close to my heart and born out of from my past is substance abuse. A year ago I helped found the SADD (Students Against Destructive Decisions) chapter at my school. I am the advisor for this group. SADD's mission is: "To provide students with the best prevention tools possible to deal with the issues of underage drinking, other drug use, risky and impaired driving, and other destructive decisions." We meet once a week and the turn-out of students has been overwhelming—I know I'm making a difference—I'll never stop trying.

Describe your personal feelings and beliefs about teaching, including your own ideas of what makes you an outstanding teacher. Describe the rewards you find in teaching (from Section V-A of the application: Philosophy of Teaching)—My educational philosophy is based on the following feelings and beliefs. I consider them foundational to my work as a teacher:

I believe that every student should be afforded an equal opportunity to learn. This requires that each student be considered as both "gifted" and of "special needs." Further, it inherently requires that I continuously monitor each student's progress and intervene, when necessary, at the earliest possibility to effectively promote student success.

I believe that every student deserves a "highly effective" teacher in every classroom. To this end, I have committed myself to being a "life-long learner"—to pursuing every opportunity for professional development so as to ensure continuous improvement of my pedagogy; and, I am committed to sharing my knowledge whenever possible with my peers and those whom I mentor. In the spring of 2011, I volunteered as a teacher to be evaluated in the pilot of the new Rhode Island Department of Education's Educator Evaluation System. This positive experience has instilled in me a belief that this system will place more highly effective teachers in the classroom and help ensure a bright future for education in Rhode Island.

I believe that a teacher's attitude should be that, given enough time, every student has the capability of learning anything and that the teacher's agenda should aim at this end. But my experience has also made me a realist, and I also acknowledge that this is unfortunately not always possible for many reasons. The "mind set" of the teacher should be always be "positive"—no child should ever be written off as un-teachable or unreachable.

I believe in "content" and "performance" standards. They establish and ensure consistency in teaching and learning independently of student demographic and/or socioeconomic status. However, I also believe that a measurable benefit of standards is the ability to efficiently and effectively target areas of student deficiency and so remedy them.

I believe that every subject should be taught with techniques and strategies that engage students and make maximal use of their learning time. I often give my students the opportunity to have a voice in their homework and project assignments. In my experience, these "flexible" assignments have been overwhelmingly successful across all levels of student abilities. Students have demonstrated deeper levels of learning, been engaged for longer periods of time, improved their self-esteem and confidence, and perhaps most importantly, found relevance in what we are all trying to do.

How are your beliefs about teaching demonstrated in your personal teaching style? (from Section V-B of the application: Philosophy of Teaching)—Every decision I make is student-centered. I come to school every day with a smile. I am respectful to my students as I want them to be to their peers and myself. I take every opportunity to support them in their learning experience. I provide multiple opportunities for them to demonstrate proficiency and also to receive the remedial assistance when necessary. I maintain a website which contains resources which help them with their academic work. I also work very closely with my grade-level team to ensure that each and every student's needs are being met.

I am constantly aware that I am a major point of focus for my students in my class. The fact that I am for them a role model demands that I take responsibility and ensure that they are always seeing a teacher who models appropriate behavior and uses appropriate language.

What do you consider to be the major issues in public education today? Address "one" of them in depth, outlining possible causes, effects and resolutions (from Section VI: Education Issues and Trends)—My list of the "top ten" major issues in education today, in no particular order, are: NCLB; teacher testing/effectiveness; teacher pay; standardized testing; school funding; unions and collective bargaining; social promotion; the public perception of the teaching profession; school choice and vouchers; and, charter schools. Of these, the topic of charter schools is of particular interest to me as I live the experience each day. I am concerned with some of the public misconceptions of charter schools. My thoughts on this issue are:

For more than two decades, charter schools have had a polarizing effect on education. Proponents of the charter school movement argue that such schools offer students an alternative to the stagnant or failing schools in their districts; they offer a compromise on the controversial question of vouchers because they often allow public money to be used for private schools. Also, many charter schools offer specialized education focusing on areas such as art, environment, math, science, and technology. Opponents argue that they effectively steal funding from local school districts, hire non-certified teachers, underpay them, underperform in comparison to their sending districts, and in many cases, "cream" off the better students in their selection process.

Problematic to this debate is the fact that that charter schools remain a mystery for most of the general public. Briefly described, charter schools provide an alternative to traditional public schools. They are self-governed and provide greater autonomy for educators to operate in a domain that is free from district rules and collective bargaining agreements. They are expected to be social and educational laboratories and to serve as vanguards in the process of improving and expanding opportunities within public education. And, since they are public schools, attendance is free of charge.

However, in exchange for the greater autonomy they possess, charter schools are held more accountable. In Rhode Island, their charter is reviewed every five years and can be revoked for non-performance.

Unfortunately, all too often the debate over charter schools lumps all of them together. Charter school laws vary dramatically from state-to-state and despite the fact that they all operate under the general framework described above, they differ dramatically from each other. Let me address a few of the common talking points raised by critics of charter schools.

Rhode Island has implemented a new school funding formula. It was developed in such a way that it is equitable for all schools. My colleagues and I are all fully certified per state law. We are paid wages that are comparable to those of teachers in the local school district. As far as student achievement is concerned, Beacon Charter High School is still the only charter school to have been the state's top performing district on a NECAP test (97 percent proficient in Reading in 2009). In 2010, we beat the state's NECAP averages in reading, writing, and math. Over the last four years, Beacon has posted the highest gains of any district in the state in student achievement on the NECAP tests in reading, writing, and math. We have proven that these goals are attainable. So, the new challenge is to make this trend sustainable. We are a U.S. News and Report's Bronze Medalist "Best High School" winner for both 2008 and 2009.

As for the claim that we "cream" the best students off of the system as a whole, all of Rhode Island's charter schools are obligated to conduct lotteries because of overwhelming demand and the lack of available seats. This I attribute to "customer satisfaction"—a metric deserving of more respect than it often gets. This year, nearly 30 percent of our incoming ninth graders have IEPs—one of the highest percentages of all high schools in the state. A significant number of them have tested at the 5th to 7th grade levels in mathematics and are two to three years below grade level in reading. And, where qualifying for "free or reduced" lunch is concerned, we have one of the highest percentages in the state.

Perhaps successful charter schools like ours (I consider myself fortunate to teach in it) need to do a better job on sharing our innovations and marketing our successes, especially in view of the fact that we typically educate a disproportionate number of "high risk" students.

What do you do to strengthen and improve the teaching profession? (from Section VII-A of the application: The Teaching Profession)—I take advantage of every professional opportunity possible and share my experience with my colleagues. Last year, I attended workshops on *Understanding by Design* (UbD); and subsequently I assumed a teacher-leader role in facilitating several professional development workshops for colleagues after school. I am a co-author of our school's RTI (Response-to-Intervention) Handbook. I recently participated in an online discussion board with nine other teachers involving exploration of RTI related issues. This past spring, I volunteered to be a participating teacher and so be evaluated in RIDE's pilot of the new Rhode Island Model Educator Evaluation System.

What is and/or what should be the basis for accountability in the teaching profession? (from Section VII-B of the application: The Teaching Profession)—I believe that every student deserves to have a highly effective teacher in every classroom. To this end, teachers should be held accountable for effectively demonstrating that "evidence-based" best practices have been fully implemented in their classrooms; that, where applicable, the appropriate standards are being addressed; and, that reasonable and measurable student learning objectives have been established and are being met.

Problematic for this lofty goal is the measurement of "teacher effectiveness" in a fair and equitable way. Accountability cannot be based solely on "one-shot" student performance on state tests or on those relatively infrequent, short snapshots of classroom activities that classroom observations amount to. These measures, arguably more quantitative than qualitative in nature, too often fall short of capturing the true essence of the continuum of teaching and learning that goes on in the classroom. It is my belief that any fair and equitable accountability model must incorporate multiple measures, preferably from different stakeholder perspectives not just that of the state's data analyst, the school's building administrator or the department head.

I believe in "transparency for accountability"—the disclosing of all the details of one's practice for purposes of seeking critique and assistance in perfecting one's pedagogy. Transparency for the sake of accountability is embedded in every aspect of my pedagogy. It begins on the first day of class with my syllabus. This is where I clearly lay out for students, parents, and administrators, the content to be taught in the course, state learning objectives and identify standards, provide grading policies, and list other expectations. I maintain a website which includes class related activities and resources for students and parents.

At Beacon, all teachers maintain a daily online "curriculum diary mapping Wiki." My postings include daily class activities, student learning objectives, and standards being addressed. The Wiki is visible to all teachers across the content areas. It makes it possible for us to plan interdisciplinary activities and know which standards are being addressed at a given time. I recently received an award from the administration for the "Best Curriculum Diary Mapping Wiki."

We use PowerSchool, a web-based grading system that allows students and parents to access grades 24/7. Posting my grades daily provides a true picture of a student's progress in real-time for both students and parents.

I frequently use student reflective essays, online discussion boards, and exit surveys to poll my students on the teaching and learning that is happening in my classroom.

Also, I often seek suggestions from my peers by sharing my lessons and or teaching strategies—there is always a better way to do things.

Prior to this year, Beacon used the "Toledo Peer Assistance and Review" model for teacher evaluation. Colleagues observed teachers and worked together to improve practices. This year, we have implemented the new Rhode Island Model Educator Evaluation System. Last spring, I was a participant in the state's pilot for this. The model involves multiple measures and I believe it has been structured to be fair and equitable. However, I don't believe the "accountability" aspect of it has been finalized yet.

As the 2012 National Teacher of the Year, you would serve as a spokesperson and representative for the entire teaching profession. What would be your message? What would you communicate to your profession and to the general public? (from Section VIII of the application: National Teacher of the Year)—Teaching and learning is one of the most complex processes in society—made complex because of the many differences between all stakeholders; also there is the reality that every aspect of education is really an invention of human thought. Teachers teach differently, students learn differently, schools and districts operate differently, states have different expectations, levels of parental support vary as does home life. Many initiatives in education appear and then quickly disappear. In recent years, where there has been greater emphasis on heightened accountability in education, the teaching profession has all too often been the "bulls-eye" on the target. We are playing a "Blame Game." The national coverage of some negative events has polarized the public perception of teachers and schools. All too often these things obscure the great teaching that goes on everyday across the country.

Examples of this include national media headlines such as "All Teachers Fired in Providence" (February 24, 2010); "Rhode Island Teachers Fired: Central Falls High School Officially Fires all Teachers" (February 18, 2011); "How Collective Bargaining Forced Milwaukee to Fire a Teacher of the Year" (March 8, 2011).

I find it extremely disheartening to type the words "teacher" [space] "student" into Google and haul up an overwhelming number of sites which discuss negative actions and events which center on inappropriate teacher-student relationships. Other contributors to the problem include the numerous negative postings on YouTube and Facebook. We need more positive movies such as *Stand and Deliver*, *Mr. Holland's Opus*, *The Ron Clark Story*, *Lean on Me*, *Dangerous Minds*, *Good Will Hunting*, and *To Sir with Love*.

As the 2012 National Teacher of the Year, it would be my mission to promote the great teaching that occurs every day across this country in the spirit of this prestigious "Teacher of the Year" program. [This ends Mrs. Waters' application essays.]

November 18, 2011 (Friday) 7:00 a.m. I arrive at school, Mike is busy arranging coverage for Mrs. Waters' classes today. Her husband, Chris, called him last night to let him know that she had been admitted to the hospital to deliver the baby. Coverage for her classes beginning next week would be no problem. We had already interviewed and hired a long-term substitute. He would start on Monday.

3:33:08 PM EST. An email is sent to the entire Beacon community:

> Colleagues,
> Jess and Chris are proud to announce the arrival of a baby girl late this morning. At 6 lbs., 1 oz., their third daughter couldn't wait for her due date! Warm wishes to the whole Waters clan!
> M

23 minutes later. Mrs. Waters forwards an email to me that she received two hours earlier from the Rhode Island Department of Education. The email was sent to all the three finalists:

> Good afternoon,
> I hope you are all well. I am amazed that we are already almost through November. I wanted to follow up to let you know that the announcement for the TOY [Teacher of the Year] will not be until after the holidays. I hope this offers you some relief to relax and enjoy the holiday season. You will be notified sometime early in January.
> Take care,
> [name deleted]

Of course, everyone would like to know who the winner is, but at least we don't have to wake up every day wondering if today is the day for Mrs. Waters to be named TOY and how would we handle the awkwardness of the Governor and Commissioner showing up at the school when she is out on maternity leave? With this distraction temporarily out of the way until early January, Mrs. Waters could now enjoy the holidays with family and friends; and of course, "Seraphina Grace Waters."

November 30, 2011 (Wednesday), 6:21 p.m. I receive a text with simply the words "The biggest loser." It was from Mrs. Waters.

I didn't think anything of it until I woke up the next morning and turned on the news around 6:00 a.m. It was shocking! There was the Governor and the Commissioner of Education awarding the Rhode Island 2012 Teacher of the Year to the finalist from Coventry. What happened to the "relax and enjoy the holiday season. You will be notified sometime early in January" in the November 18 email?

It was a nice surprise for the winner regardless. Jessica informed me that Julie Lima Boyle is a deserving candidate and she will be a great representative for the state.

By the way Mrs. Waters, I'm thinking of making a trip to Foxwoods if you care to join me . . . I have a system!

THE PETER MCWALTERS AWARD

December 14, 2011. Principal Mike Skeldon calls me into his office and tells me there has been a request made to all charter school directors to nominate one educator from each school for the *Peter McWalters Professional Educator Award.* The award, including a $1000 stipend, would be presented at the 2012 Rhode Island Public Charter School Consortium to be held Tuesday, March 13, 2012 at the Warwick campus of CCRI (Community College of Rhode Island).

This would be the second year for the award, named after Peter McWalters who served as the Rhode Island commissioner of elementary and secondary education from 1992 to 2009. Commissioner McWalters is credited with several accomplishments toward improving Rhode Island's public schools including promoting "choice quality education."

The award application instructions that Mike has handed me include the selection criteria that states the nomination should be based on the educator having: (1) contributed in measurable ways to the improvement of the school; (2) demonstrated exceptional dedication and skill in shaping and influencing children's lives; (3) exhibited innovation in teaching methods; (4) inspired and encouraged students of all backgrounds and abilities in their learning; and, (5) earned the respect as a positive role model and professional by other teachers and colleagues.

The three member selection committee will be comprised of Denise Jenkins of the Rhode Island Foundation; Jennifer Wood, Esq., Chief of Staff/ General Counsel, Office of Lieutenant Governor; and, Kate Nagle from the RDW Group.

At this point, I was thinking "isn't this great—that Mike is nominating me for this prestigious award!" In two seconds, I had already spent the $1000 stipend. Unfortunately, somewhere between the third and fourth seconds, I had to quickly reel back my thoughts as I heard Mike's next words.

He tells me he would like to nominate Mrs. Waters for the award and, since I have a wealth of inside information on her (i.e., my research), he would like me to write the required one page nomination narrative stating why she is worthy of the honor. He would approve it and then sign it. However, Mike asks me, and I agree, not to tell her because we don't want to provide a second disappointment if she doesn't win. The nomination narrative is due at the Rhode Island League of Charter School's office by January 15.

I go back to my office and begin working on the nomination letter. My biggest challenge would be to try to address all five of the selection criteria and fit this information onto one page. I decide to frame the narrative in the order of the five criteria topics and just write—I would condense it later. I made good progress before calling it a day and heading home. Later, that evening, I continued working on it. When I had addressed all five criteria, I had filled five pages—only missed the mark by four pages. What could be cut? I spent the next day trimming, condensing and compressing so as to finally reduce my text to the required single page. Of course, I had to use half-inch margins all around and reduce the Cambria font to a 10-point size in order to fit my 921 words (66 lines) to one page:

Nomination Narrative for Jessica Walters, Beacon Charter High School

Dear Selection Committee,

I am submitting the name of Jessica Waters, lead science teacher at Beacon Charter High School, for this prestigious educator award. Jessica graduated from Rhode Island College in 2008 earning a Bachelor of Arts Degree in Secondary Education with a biology major and chemistry minor. She joined our faculty in the 2008-2009 school year and is currently in her fourth year of teaching. She has taught physics, chemistry, human anatomy & physiology, and SAT Prep. I offer the following narrative addressing the Peter McWalters Professional Educator Award's five selection criteria areas:

Contributing to measurable ways of improving the school: Jessica Waters is the teacher of record directly responsible for doubling Beacon's NECAP science scores and beating the State's average scores in all four science categories; Beacon administers the NWEA assessments twice a year. Jessica's science students continue to surpass the NWEA national growth norms; She is a co-author of the school's Response-to-Intervention (RTI) Handbook; Beacon was one of only three RI schools chosen to pilot RIDE's new educator evaluation system, referred to as The Rhode Model, in the spring of 2011. She volunteered to be a teacher who would be evaluated—her input has been invaluable to both the state and our school; and, she volunteered to be observed for 79 consecutive days in her chemistry class for a research project studying classroom management and student engagement and is the focus of a book entitled *Who's the New Kid in Chemistry Class? Exploring Uncharted Waters* being released for publication this spring; Jessica has conducted professional development workshops for the faculty on Understanding by Design (UBD)—a subject on which she has become an expert. As lead teacher, she is currently mentoring a new science teacher who recently joined our faculty; also, she has influenced Beacon's "Guiding Principles" on homework, projects, and classroom management.

Demonstrating exceptional dedication and skill in shaping and influencing children's lives: Jessica has a very interesting life story, one that serves as an inspiration to all of our students. Her early years were not unlike those of many of the students she teaches. She grew up in a low-income home with drug addicted parents. When she was in 10th grade, she witnessed her parents being arrested and her little sister running into the woods out of fear. Subsequently, Jessica dropped out of school, worked two jobs, earned her GED, and enrolled in CCRI. She wanted to be a doctor someday, but later decided that, she would become a caring teacher who could make a difference in somebody's life. Upon joining Beacon, and being influenced by her past and her desire to positively shape children's lives, she co-founded and serves as advisor for our SADD (Students Against Destructive Decisions) chapter. SADD's mission is: "To provide students with the best prevention tools possible to deal with the issues of underage drinking, other drug use, risky and impaired driving, and other destructive decisions." Jessica also has an advisory which has won Beacon's Annual Thanksgiving Food Drive three years in a row—another testimony to her dedication to shaping and influencing children's lives. Perhaps most impressive of all is the increasing number of Beacons students who have expressed a desire to pursue careers in science-related fields (e.g., medical)—quite an accomplishment considering Beacon's intense focus on the arts.

Exhibiting innovation in teaching methods: At Beacon, all teachers maintain a daily online "curriculum diary mapping Wiki." Postings include daily class activities, student learning objectives, and standards being addressed. The Wiki is visible to all teachers across the content areas. It makes it possible for us to plan interdisciplinary activities and identify the standards that are being addressed. Jessica recently received an award from Beacon's administration for the "Best Curriculum Diary Mapping Wiki." She frequently uses student reflective essays, online discussion boards, and exit surveys to poll her students on the teaching and learning that occurs in her classroom. This is reflective of her commitment to continuous improvement of pedagogy. Jessica has also pioneered "Flexible" homework and projects in which students have a voice in creating their own assignments. The aim of this is to tap student interest and promote engagement. She has also pioneered methods for differentiating summer reading assignments based on NWEA science assessment areas of student weaknesses.

Inspiring and encouraging students of all backgrounds and abilities to learn: As a member of Beacon's RTI team, Jessica has been actively engaged in ensuring that students of all abilities succeed. She has taken a leadership role in grade-level meetings and department meetings in which student performance and interventions are discussed. I recall a recent case in which a low performing, special needs student who was suffering from low self-esteem, produced a very impressive chemistry project. Jessica encouraged her to present her project in two other chemistry classes. The student agreed and did an exceptional job, freeing herself from her low self-esteem. Jessica constantly offers extra-credit opportunities to challenge learners at all levels.

Being respected as a positive role model and professional by other teachers and colleagues. Jessica is well respected by her peers, as evidenced by her nomination by our faculty this year for the Rhode Island 2012 Teacher of the Year. Although she did not win, we are extremely proud of her being selected by RIDE as one of the three finalists for the award. This was the first time a charter school teacher had made it to the final stage—a milestone event that was neglected by the media.

Unfortunately, the one-page limitation for this narrative has only allowed me to briefly highlight some of the key reasons why Jessica is most deserving of this honor. I hope you will give serious consider to her nomination.

Sincerely,
Michael Skeldon, Principal,
Beacon Charter High School
[end of narrative]

Mike found a couple of grammatical errors in the text, corrected them and signed the nomination narrative and emailed the application into the committee three days in advance of the deadline. Now we would have to wait two months to find out if Mrs. Waters would win. To the best of my knowledge, Mike and I are the only ones in Beacon who know that she has been nominated for the award.

February 1, 2012. No word on the McWalters Award but I receive an interesting telephone call from Mrs. Waters. She tells me that RIDE called her earlier today about Rhode Island's new expanded Teacher of the Year Program. Because there is a new timeline and transition for the program effective 2012-13, she is actually still in the running and will be a finalist for the 2013 process. Briefly, here is what she was told about the new process:

Each district in Rhode Island will select and submit the name of their district Teacher of the Year (DTOY) to RIDE by mid-April. RIDE provides a year of support for the DTOYs in preparation for completing the application for the state Teacher of the Year (TOY). RIDE's selection panel reviews all of the prior year DTOY applications, chooses the top six, and then conducts on-site observations in May. From these six observations, three finalists are chosen for interviews in June. After the interviews, the panel selects the state Teacher of the Year. The award is announced at a ceremony in September. RIDE submits the state TOY as Rhode Island's candidate for the National TOY process beginning in November.

Mrs. Waters tells me that she needs to resubmit her application by March 26. She was told she could send in the same essay responses written in her original application (those cited earlier in this chapter) and that she needs a "Letter of Support" from me. Also, RIDE's search panel will come to Beacon in May to observe her in the classroom. This will be a little awkward as Mrs. Waters is out on maternity leave and will have to come back for only one day in May for the classroom observation and an interview.

Mrs. Waters confides in me that she is not sure she wants to continue with this process because of all of the stress it is causing. However, before we end our telephone conversation, she agrees to do it. I'm thinking that not only does she have a second chance at Teacher of the Year, but unbeknownst to her, a chance at the Peter McWalters award as well.

February 9, 2012. RIDE calls Principal Skeldon to confirm that Mrs. Waters is indeed one of three finalists for the 2013 state TOY and explains the new expanded process exactly as Mrs. Waters described it to me. It allows the three state TOY finalists to hone their presentation and interviewing skills, particularly for the "mock press conference" component in which Mrs. Waters felt she had bombed her question on NCLB.

March 12, 2012 (the day before the 4th Annual Public Charter School Consortium). Mike informs me that Steve Nardelli, President of the Rhode Island League of Charter Schools, had called to inform him that Mrs. Waters would receive the Peter McWalters Award tomorrow night at the Consortium. He was also calling to check the spelling of Mrs. Waters' last name for the award plaque. Apparently, in the process of writing her nomination application, I spelled "Waters" as "McWaters!"

THE 4TH ANNUAL PUBLIC CHARTER SCHOOL CONSORTIUM

March 13, 2012 (Warwick campus of the Community College of Rhode Island). I arrive at the college (site of the Consortium) about 3:00 p.m. Several weeks ago, I had pre-registered for the event but I wanted to get here early to get a seat as close to the podium as possible. The award ceremony was scheduled for 4:00–4:30 p.m. Dr. Henry Fuller, distinguished Professor of Education and director of the Institute for the Transformation of Learning at Marquette University would present the keynote address from 4:25–4:55 p.m. Breakout sessions would be conducted between 5:00 and 7:30 p.m. followed by coffee, dessert, and a raffle for an iPad. I proceed to the registration desk where I was handed a folder with a label that had my name on it. I then proceeded to the stage and podium at the far end of the large open cafeteria area. Evidently, I am one of the first arrivals, so finding a table directly in front of the podium was no problem. I save the end-seat for Mrs. Waters, figuring she'll have to get up to go to the podium, and then I claim the entire table for the several Beacon faculty members expected to attend in her support. This venue is a sort of déjà vu for me. I spent two years in this cafeteria while attending the college for an Associate in Science in Engineering degree between 1975 and 1977.

Mrs. Waters arrives, pushing four-month old Seraphina in a stroller accompanied by 11-year old daughter Mackenzie. She will watch Seraphina during the presentation. Chris is home taking care of their middle daughter, Madeline.

More Beacon faculty now arrive: Mike, Carolyn, Pete, James, Carrie, Meghan, Sarah, Ann, Jill, Shayna, even former Principal Dr. Pilkington.

We grab some refreshments and then take our seats as the event begins. Jennifer Wood, one of the search committee members, and Chief of Staff/General Counsel from the Office of Lieutenant Governor provides an introduction to the award and comments on why the committee selected Mrs. Waters. Interesting enough, she quoted many of the words I had written in the nomination narrative. She then calls "Mrs. McWaters" to the podium (Oh no!) to receive the award. It consisted of an engraved plaque and a $1000 check. Mrs. Waters returns to her seat next to me. I check out the plaque to see how they spelled the name. Great! It is correct (Waters not McWaters). I ask her if I can see the check. It is indeed made out for $1000. I slip it into her folder on the table between us.

Dr. Howard Fuller delivers a very inspiring keynote address. We then proceed to breakout sessions for the next two-and-a-half hours. I attend one on RTI because our faculty member, James MacDonald, is a presenter. I also attend one session on science.

The Butler Did It

During the break between sessions I and II, I return to our table in the cafeteria. As I approach the table, Mrs. Waters stops me: "Okay where is it." I ask her what is she talking about, "My check," she says. I respond "I don't have your check." I am now getting a sick feeling in my stomach. This is very embarrassing—the recipient of the prestigious Peter McWalters

Professional Educator Award has lost a $1000 check. Everyone is looking at me, expecting me to admit that I have been pulling off a prank. “I know it was either you or Pete,” says Mrs. Waters. I’m thinking it’s Pete too because he is known to pull off a few pranks from now and then. Pete comes by and we ask him, but he knows nothing about the missing check. Wow, this is getting ugly! Should we inform the event coordinators that the check is missing and have them make an embarrassing public announcement? Before making this drastic move, I second guess myself and decide to go through the paperwork in my folder. I open the folder in front of everyone and sure enough, there’s the check tucked into the left pocket. Apparently, I had mistakenly placed the check in my own folder rather than Mrs. Waters’! But it was an honest mistake. The folders are exactly the same color and have the same style labels on the front. I didn’t have on my reading glasses, so it was impossible to make out the name on the folder label. Now, everyone thinks I did this on purpose. I spent the remainder of the evening trying to convince everyone that it was not some sick joke.

THE CLIMB TO STATE TOY CONTINUES

March 26, 2012 (Monday). Mrs. Waters delivers her re-application for the state TOY with my “Letter of Support” to RIDE so as to meet the deadline.

May 10, 2012 (Thursday). RIDE’s search panel comes to observe Mrs. Waters. She has chosen to perform a lab with a chemistry class. She later tells me she thought it went well.

May 14, 2012 (Monday) 3:25:45 PM EDT. Mrs. Waters receives an email from the RIDE search panel entitled “Next Steps.” It has been sent to the state TOY finalists who were just observed in their classrooms. The following are excerpts from the email:

> Good Afternoon,
>
> Thank you for opening your classroom doors and welcoming us in to observe you and your students. The classroom visit and small group interview were impressive.
>
> I wanted to share the next steps in the process so you may begin to prepare. We will be conducting Final Panel Interviews on Wednesday, May 23 from 4-6 p.m.
>
> The process will have 3 parts:
>
> a) 10-minute presentation on: “A current educational topic that interests you both philosophically and/or personally as an educator.” You may use posters, charts etc., or create a power point presentation for the panel. But your presentation is limited to 10 minutes in length.

b) Bring a single artifact that was created as a result of a lesson or created by you for a lesson. Explain its significance and impact on students and student learning.

c) Whole Group Mock Panel Interview: You will be asked a series of educational questions that pertain to your community, state and/or nation. This will be an opportunity for the panel to assess your knowledge of current educational issues and your ability to articulate your perspective and understanding of each topic.

May 23, 2012 (Wednesday). Mrs. Waters is interviewed at RIDE. She presents her PowerPoint on student engagement (slightly modified from the original). She brought Robert's flexible project painting, "The Life Cycle of a Star," as an example of student work. And, she participates in the whole group mock panel interview (i.e., mock press conference).

June 19, 2012 (Tuesday). It's Beacon's 180th school day and the last one before the summer break. Now we will have to wait for about three months (September, 2012) to hear whether Mrs. Waters is selected as the state TOY. The good thing is that she has already been recognized as an outstanding teacher by both the TOY process and by receiving the 2012 Peter McWalters Professional Educator Award.

What Does Husband Chris Think?

Q: What did you think of Jessica's recognition for TOY and the McWalters award?

A: I am so proud of my loving wife's success. Jessica inspires her students to achieve to their potential and, at the same time, gains the respect of both colleagues and parents. She deserves the recognition!

September 19, 2012 (Wednesday), 9:30 p.m. I'm in a somewhat cranky mood this morning. I have to attend a professional development seminar on "data usage" tomorrow, Friday, and next Tuesday. So, I have to prepare the class work and homework assignments for the substitute teachers for the classes I will miss. Being out of the classroom is troublesome indeed.

Principal Skeldon stops by my office. The state's science scores have just been released to the schools and will be made public tomorrow. He beams as he tells me that we increased our percent proficient from 28 percent to 45 percent; the state average is 32 percent. The net increase in our score over the last two years is therefore 31 percent—a solid accomplishment.

Next, Mike asks:

"John, are you going this afternoon."

(Mike is referring to the "Rhode Island Educator Recognition Celebration." It takes place later this afternoon. The state will announce the winner of the Rhode Island 2013 Teacher of the Year and also recognize the new district Teachers of the Year (DTOYs). Pete Boland, Beacon's English teacher is our new DTOY.

"I wasn't planning on it."

"You really should go, you need to support your colleagues."

"Mike, I wasn't planning on going. I thought it was just a principal–teacher thing. I've got to get the next two days worth of assignments ready for the students, and also I have the grade-level meeting after school. I really dislike these events, and if Jessica doesn't win, I will be really depressed."

"John, you need to go."

Mike leaves me feeling guilty. I think about it for a few minutes and then look for him so as to get the address and the time of the event. He tells me that light refreshments will be available from 4:00 to 5:00 p.m. and the actual ceremony will begin at 5:00. Several awards will be handed out, culminating in the major announcement of the Rhode Island 2013 Teacher of the Year.

5:30 p.m. I arrive at the Orchard Farms Elementary School in Cranston, the site of the ceremony. I contemplate just sitting in the car and texting Mike or Mrs. Waters to see if she has won, but then decide to head in. The auditorium is filled and someone is speaking from the podium in the front. I decide not to make a scene by looking for my Beacon colleagues, so I look for a seat in the back row. Unfortunately, as I approach an empty seat on the end of the back row, I trip over an empty water bottle somebody left on the floor. It takes flight, banging against the wall twenty feet away. I dive for the seat. Several heads turned to see what the commotion was—I immediately point to the person sitting next to me. Needless to say, she was not happy.

The stage is filled with various dignitaries. I recognize Governor Lincoln D. Chafee. They are still handing out awards so I haven't missed the announcement. I spot Mike and some of Beacon faculty several rows directly in front of me. He keeps looking around but doesn't stretch his neck far enough to see me directly in back of him. I assume he is trying to locate me and decide to text him to let him know I'm here. I can tell by his body language that he got the message. I see that several of Beacon's faculty are sitting with him as well: Pete Boland, Tina Go, Dana Pagliaro, Carrie Appel, and Robin Murphy, and possibly others that I cannot see. Husband Chris and the three daughters, 12-year old Mackenzie, 3-year old Madeline, and 10-month old Seraphina (in a baby carrier) are sitting with the group.

5:50 p.m. The Governor approaches the podium to make the announcement. He begins his speech by describing the selection of the Teacher of the Year, but coyly avoids mentioning the winners name by saying "He or She." Nervous energy traverses my entire body. His speech seems like it lasts an hour. "He or She" is a science teacher. Good! Finally we have a clue. Then he gives us the name:

"The Rhode Island 2013 Teacher of the Year is Jessica Waters." Well, all of the months of thinking about this—virtually every day—are finally over. She won! She receives several gifts including a $10,000 scholarship from Johnson & Wales University that can be applied toward their doctoral program in educational leadership, a free trip to a national park, and $2500 for school supplies.

Mrs. Waters spends the next hour talking to the television crews and other members of the media who are present for the ceremony.

I stick around with my Beacon colleagues out in the school's main lobby, waiting to congratulate Mrs. Waters. Mike tells me that he had received a call at 9:00 this morning from RIDE informing him that she had won. He swore the cabinet (Carrie Appel, Tina Go, Pat Hawkridge, Robin Murphy, Carolyn Taylor) to secrecy—they were to tell no one, especially me (no respect for me keeping a secret). That's why he stopped by my office to tell me I should go and also why he was doing neck stretches looking for me in the audience.

Meanwhile, husband Chris is being interviewed by a reporter, so I end up entertaining their 10-month old daughter Seraphina. I'm tickling her little feet and making obnoxious noises with the back of my hand pressed against my mouth. She's laughs. When I stop she cries. So, I have to keep tickling her and making funky sounds.

Finally, Mrs. Waters appears and we all give her a hug and congratulate her. This is certainly going to be a life-changing event for both her and her family. And, it is obvious that it is also an important occurrence for the entire Beacon Charter High School community. Mike takes out his iPhone and shows me a picture of a huge banner he has designed for the front of the school:

Home of Rhode Island 2013 Teacher of the Year
Jessica Waters

I'll sleep really well tonight!

A NEW BEGINNING

In the months that immediately followed, Mrs. Waters' life dramatically changed—invitations to speak at conferences and events, invitations to join committees, and opportunities to be introduced to some pretty amazing people (see figure 9.2 and 9.3). But this is just the beginning.

Figure 9.2. Mrs. Waters and Arne Duncan, US Secretary of Education. (Photo courtesy of Ron Sachs, CNP.)

Figure 9.3. Mrs. Waters and Dr. Jill Biden. (Photo courtesy of Ron Sachs, CNP.)

NOTE

1. *Field of Dreams* was a 1989 fictional film (nominated for three academy awards) about a novice farmer, played by actor Kevin Costner, who hears a voice while walking in his cornfield. It whispers to him "If you build it, they will come." He subsequently plows under his cornfield and builds a baseball diamond. Eventually, after facing possible financial ruin, baseball players (deceased) arrive to play. If Mrs. Waters, or any of us, can make our classes lively and engaging, students will come to school—and be less likely to drop out.

Chapter Ten

Reflections on the Research

> Truth is one forever absolute, but opinion is truth filtered through the moods, the blood, the disposition of the spectator.
>
> —Wendell Phillips[1]

Writing this book has often jogged my memory and freed up some of the details of many hitherto forgotten events. Preparing this chapter was as case in point.

Back in 1979, my senior year at Western New England College, our cohort of engineering students was preparing to make our senior project presentations before a panel of judges. The winner would represent the college in a New England engineering competition involving a number of schools including the universities of Connecticut, Maine, Massachusetts, New Hampshire, Rhode Island, and Vermont; several other colleges were included as well, e.g., Dartmouth College and Worcester Polytechnic Institute.

The preparation for this involved audio/video recording of several practice rounds of each of our presentations. Peers and some faculty members acted as a mock panel, providing a constructive critique of our performances. It was amazing to hear our voices on tape or see images of ourselves on film. They differ considerably from our conscious 'takes' on ourselves. I remember one student who had a habit of uttering the word "uh" before every sentence—an extremely annoying habit from the audience's perspective, and he never realized he was doing this. Another student had a habit of never looking at the audience—his head was always tilted down as he read his presentation from index cards. As for myself, I remember I was criticized for talking and indexing my slides too fast—amongst other things. When I viewed the tape of my first practice round, I found that my mannerisms, body language, and some voice patterns distracted from the content of the presentation. Listening to myself was akin to torture! Needless to say, I found several areas in which change was necessary. So, through this peer-reviewed and self-reflective process, we improved upon our final presentation. For me, putting aside my ego and listening to feedback from others, together with self-reflection, eventually paid off. I won the school competition, went on to win the New England Regional event, and then participated in two national competitions, one in New York and one in St. Louis. Placing second in the latter landed me a great job offer, and my senior project led to the development of two US patents.

Over the years, I have had an opportunity to mentor many people. On a few occasions, I have had to subtly admonish some of them for acting like know-it-alls. I would tell them that there is a difference between knowing a lot and being a know-it-all. The former requires that you listen to others, reflect, investigate, research, explore, etc.—then you have data to act upon and can work toward continuous improvement. In contrast to this, the *know-it-all*, has no need to listen to suggestions, seek feedback, or be self-reflective.

From my first year of teaching onward, I have required that students write reflections on their math assignments. They are instructed not only to reflect on their experiences but also on my teaching. I ask them, for example, whether or not I explained the lesson clearly, whether they found the assignment too easy, whether I need to slow down, whether the class was engaging or boring, etc. From their reflections and comments, I am able to quickly assess the effectiveness of my teaching for that particular assignment and determine whether corrective action is warranted. Over the years, I have gained great appreciation for the need to self-reflect and continuously solicit feedback from students, peers, and participants in my research. In the course of their schooling, students spend a lot of time in classrooms. They are exposed to a wide variety of teaching styles and types of subject matter, so they can explain in great detail the extent to which they are either tortured by it all or engaged in the learning process.

To this same end, in addition to addressing the validity of my research, participants were asked to reflect upon their experience of my presence for 79 days in their classroom. When I began the research, there were 17 students (excluding myself) in Mrs. Waters' Chemistry Block 1 class. During the course of the semester, two students were transferred to a lower performing class and a third was off campus taking courses at a local community college. Fourteen responses were therefore possible. Students were not required to respond. I received 10 reflections from them for a 71 percent response rate. The four students who did not respond had said they were going to submit responses but never did, in other words, nobody refused to write the reflections. I did not follow up on the missing four since I wanted reflections that were offered voluntarily, not forced in any way.

The student reflection request contained three prompts:

1. Did Dr. Butler's presence in your class interfere in any way with your learning? If so, please explain.
2. Did you consider Dr. Butler as a fellow student in the class?
3. Did you feel you had to change your behavior at all?

I also asked Mrs. Waters if she could submit a reflection on my research in her class. I framed her reflection with the following two prompts:

1. How did you feel before, during, and after the research?
2. What did you learn about the students or your teaching that was of significance, if anything?

Below are the responses from Mrs. Waters. They are followed by those from the 10 students.

MRS. WATERS' REFLECTION ON THE RESEARCH

1. *How did you feel before, during, and after the research?*
    - *Before*: I was very nervous. I felt a little intimidated to have another teacher watching me every day. I thought it was going to make me feel as if I was being evaluated every day and I was going to have to do "special" lessons because someone was watching me. I was not sure how the kids would respond to John. I also did not think it would last the whole semester.
    - *During*: After a few days I realized that I would have to just be myself because a "show" could not last very long. Soon after the semester began, John just became a "student" and, to my amazement, blended in with students. After I realized that John was in it for the long run, I began to see the great benefits from this research. I did not feel intimidated but rather John and I became a team. John was able to give me a perspective that I would otherwise never see. John gave me great advice looking through the lens of a student.
    - *After*: I felt so lucky to have participated in this study. The wealth of information that will be generated from this is priceless. I could not believe that John lasted the whole semester. Students actually viewed John as a peer. It was incredible. In fact, I think this should be made into a movie!! I think Kim Kardashian can play me.
2. *What were the significant things you learned about the students or your teaching (if any)?*

I have a greater appreciation for the use of data to drive instruction. The whole semester was one giant data set with small subsets built in. I learned to use student surveys to help improve instruction. The "flexible homework/project" was born. I learned many organizational techniques (i.e. templates for problems and chemical of the day). I developed a great appreciation for time on task and the amount of time that gets eaten up for things like field trips, school wide testing, kids pulled for art related tasks, announcements, and various other interruptions. I also became more confident in my teaching because John made it clear to me that I was indeed doing a good job. Before this I never had conformation that I was doing anything right or that others would see value in skills that I could bring to the table.

MANAR'S REFLECTION ON THE RESEARCH

1. *Did Dr. Butler's presence in your class interfere in any way with your learning? If so, please explain.*
   When Dr. Butler was present in our chemistry class, he didn't really interfere in a negative way. Instead, sometimes he'd bring up questions that really got the class thinking. Sometimes even questions that Mrs. Waters didn't know the answers to, that way she'd ask us to answer them for the next class to earn bonus points. If anything, Dr. Butler's interference was positive than anything negative.
2. *Did you consider Dr. Butler as a fellow student in the class?*
   Yes, I did consider Dr. Butler as a fellow student because he sat down and took notes or paid attention in class just like any other student. But the only thing I couldn't do to treat him like a fellow student, was to call him by his first name. Even though he was being a student, he was still an adult and one of my former teachers.
3. *Did you feel you had to change your behavior at all?*
   I didn't feel like I had to change my behavior while Dr. Butler was in class. I acted just as I would if he wasn't there or if I was in any other class. Dr. Butler sat three seats away from me anyway so either way, where he was, didn't affect the way I acted at all.

BRIAN'S REFLECTION ON THE RESEARCH

1. *Did Dr. Butler's presence in your class interfere in any way with your learning? If so, please explain.*

Dr. Butler made no interferences with our class whatsoever and at times made it more fun and educational for the class as a whole.

2. *Did you consider Dr. Butler as a fellow student in the class?*
   Yes, I did consider him a fellow student, perhaps a tad mature (ha ha), but nonetheless a fellow student that we socialized with.
3. *Did you feel you had to change your behavior at all?*
   Not at all, I acted the same way I would normally and felt he was fun and enjoyed his presence in class.

BRITTANY'S REFLECTION ON THE RESEARCH

1. *Did Dr. Butler's presence in your class interfere in any way with your learning? If so, please explain.*
   Dr. Butler's presence in the class did not negatively interfere with my learning in the class. However, his presence did add to the class discussions and experiments. His ideas brought up in discussions helped me generate thoughts about his ideas.
2. *Did you consider Dr. Butler as a fellow student in the class?*
   I did consider Dr. Butler as a fellow student because in no way was he treated differently. He was forced to go through PowerPoint notes and take the tests. He was even called by his first name like everyone else.
3. *Did you feel you had to change your behavior at all?*
   I did not have to change my behavior at all and even if I did on the first day of the experiment, I would not have been able to keep it up for the whole semester.

SIMA'S REFLECTION ON THE RESEARCH

1. *Did Dr. Butler's presence in your class interfere in any way with your learning? If so, please explain.*
   Dr. Butler's presence in the class actually did interfere with my learning experience, all in positive ways. His questions to Mrs. Waters' lessons always made me think a little more than if he had not been there. I found myself questioning the lesson concepts more, which enabled me to understand it better on the whole. Dr. Butler's presence also motivated me to perform better when it came to projects because he was so dedicated to them, as if he was an actual student. Overall, his being in the class pushed me to work harder and better in the course.

2. *Did you consider Dr. Butler as a fellow student in the class?*
   Considering Dr. Butler as a fellow student in class depended on the activity at hand. Since we sat next to each other, it was easy to work with him on worksheets and packets as if he was a peer. However, that wasn't the case when it came to class presentations for projects and such. The fact that he is in fact a real teacher affected his style when he presented his work, and so listening to that, I related him more to a teacher than a student.
3. *Did you feel you had to change your behavior at all?*
   I've always acted as myself towards people, no matter who it is. So I did not really feel the need to change my behavior when Dr. Butler joined our class. In general, I'm quiet anyway, so there wasn't really a need to stop talking or change my ways when he stepped in.

ROSABEL'S REFLECTION ON THE RESEARCH

1. *Did Dr. Butler's presence in your class interfere in any way with your learning? If so, please explain.*
   Not at all. In fact it was fun to have him in class because he always had interesting things to say and often was very humorous.
2. *Did you consider Dr. Butler as a fellow student in the class?*
   Yes, though he did act professionally and get his work/research done, he was very much a classmate to us all.
3. *Did you feel you had to change your behavior at all?*
   No. Classes were always comfortable and that did not change when he arrived in our class. He joked and enjoyed his time in the classroom and we were happy to have him there.

SARAH'S REFLECTION ON THE RESEARCH

1. *Did Dr. Butler's presence in your class interfere in any way with your learning? If so, please explain.*
   When Dr. Butler first told our class that he was going to be a student for the semester, I thought he was joking. We acted with him like we normally did in any other situation, thinking he was just assisting Mrs. Waters in class. After realizing that he's been coming in every day, taking notes and doing projects with us, I realized that he was being serious about being a student in class. Dr. Butler sat two seats away from me and I sometimes forgot he was there because he barely spoke and was never distracting. He

would work when we worked, chat when we chatted, and asked questions just like any other student.

2. *Did you consider Dr. Butler as a fellow student in the class?*
   At first, it took me about a week to adapt to Dr. Butler being an actual student in the class. However, by the end of the year, we were calling him by his first name John and we completely recognized him simply as a fellow student. He worked with me in group projects several times and we discussed and answered questions like other students.
3. *Did you feel you had to change your behavior at all?*
   Even before Dr. Butler started working on this experiment, I was very comfortable with him as a teacher and he was one of my favorites. So, when the experiment started, I was already fine with acting the way I normally do around him and did not change myself in any way. He was a pleasure to have in a class and I'm sure he learned from the class as much as the class learned from him.

SIRAJ'S REFLECTION ON THE RESEARCH

1. *Did Dr. Butler's presence in your class interfere in any way with your learning? If so, please explain.*
   Entering Beacon Charter High School as a ninth grader, I was very enthusiastic to meet my mathematics and science teachers. The first math class I ever took was with Dr. Butler. I remember myself being always excited to learn more from him comparing to any other teacher because of his teaching skills that opened my mind to new ways of learning. Over the years I've developed a strong sense of respect for Dr. Butler and for everything that he has done to help me. So, having him in class definitely encouraged me to do my best, even though science was already a great interest of mine. Both Dr. Butler and Mrs. Waters are huge role models for me, and so I felt very inspired to do everything to my best effort.
2. *Did you consider Dr. Butler as a fellow student in the class?*
   At the beginning when I first found out Dr. Butler was going to be in class with us, it felt unnatural because he is such an important figure at Beacon Charter High School. However, after just about a week, he became part of the class and he was just another student in class.
3. *Did you feel you had to change your behavior at all?*
   Just like I said earlier at the beginning, everyone was acting more studious. However, Beacon Charter High School students are always respectful and caring. So, eventually we all felt right at home even though we had two teachers in class.

LINDSAY'S REFLECTION ON THE RESEARCH

1. *Did Dr. Butler's presence in your class interfere in any way with your learning? If so, please explain.*
   I felt comfortable because he never acted like authority. We worked in groups with him and fooled around with him as well. We called him John which added to his presence of being a student.
2. *Did you consider Dr. Butler as a fellow student in the class?*
   Yes! Definitely
3. *Did you feel you had to change your behavior at all?*
   Not really, no my friend Jessica [Jessica G.] and I conducted a couple of experiments with him in our group. Our behavior was relatively the same with him present.

MARY'S REFLECTION ON THE RESEARCH

1. *Did Dr. Butler's presence in your class interfere in any way with your learning? If so, please explain.*
   No, not really. I acted as if he was any other student in the class.
2. *Did you consider Dr. Butler as a fellow student in the class?*
   Yes, as stated previously, I treated him as a fellow student.
3. *Did you feel you had to change your behavior at all?*
   No, not really, he was just an older one of us, we had fun.

RUBY'S REFLECTION ON THE RESEARCH

1. *Did Dr. Butler's presence in your class interfere in any way with your learning? If so, please explain.*
   No.
2. *Did you consider Dr. Butler as a fellow student in the class?*
   No.
3. *Did you feel you had to change your behavior at all?*
   No.

SOME CLOSING REMARKS FOR THIS CHAPTER

As the reader can see, the responses varied from several sentences to Ruby's one word answers. I never grade the content of reflections. As far as I am concerned, there are no right or wrong answers to such reflective surveys. Nor do I defend or contest any of the responses, but rather leave interpretation to the reader.

NOTE

1. Wendell Phillips (1811–1884). American abolitionist, advocate for Native Americans, orator, and lawyer. Cited in *A handbook of English Literature* by Francis Henry Underwood (1889, 358).

Epilogue

What Have We Learned and What Next?

> A few observations and much reasoning lead to error; many observations and a little reasoning to truth.
>
> —Alexis Carrel[1]

In deference to Alexis Carrel, this epilogue will be fairly brief.

The preface of this work discussed Philip Jackson's groundbreaking work *Life in Classrooms*. Jackson's stated goal was to remain neutral, not to attempt to answer questions such as: "How can one reach those students who have all but given up on school?" "How should teachers be evaluated?" or, "How can we help today's youngsters develop a love of science and mathematics?" (Jackson 1990, xv). But these were precisely the questions I hoped to explore through my participant observation research in Mrs. Waters' Block 1 Chemistry class. Unlike dissertation research that could be crafted and meticulously choreographed by directly focusing on these questions, I would have to capture the phenomena largely by means of observation. Maintaining an awareness of the affect I was having on the behaviors of students and Mrs. Waters, in so far as this would contribute to answering these questions, would be a delicate tightrope walk. In my participatory role, I could not compromise the students' perception that I was one of them.

Chapter 10, "Reflecting on the Research," included comments and observations offered by Mrs. Waters and the 10 students who participated in the research. At this point, I would like to look back on what I have said and sieve out what has been learned from this research. Doing this will shed some light on the problem of where we might go from here.

SOME NOTES FROM CHAPTER 1: THE NATIVES AND THE VILLAGE

Most educational research of this nature begins by setting the scene, not unlike what is done in developing a screenplay. Student demographics are investigated and summarized and the physical characteristics of the learning environment are outlined. Collecting and thinking on such data can lead to pattern recognition and/or help explain the cause and effect of behaviors observed during this exploratory learning experience.

Knowing which art discipline each student was involved with led me to realize that they often chose to draw on their artistic skills in flexible homework and flexible project assignments. But also I recall visual art students Siraj and Rosabel nicely illustrating the lab procedure with pictures of beakers and other equipment on the Pre-Lab Activity worksheets (see ch. 4) rather than the traditional flowcharting method of using common geometric shapes such as rectangles, triangles, and circles.

The fact that there were 16 females and only one male student eliminated any need to work on the question of gender differences as concerned student engagement in Mrs. Waters' Block 1 Chemistry class. However, in chapter 4, "Mrs. Waters' Formal Lab Report," my analyses did cover her other two chemistry classes. These allowed for consideration of the role of gender because of the mix of students. No significant statistical differences were found between females and males as to performance on the formal lab report assignment.

With regard to SES (socioeconomic status), the cohort was relatively evenly split: 47 percent qualified for free lunch or lunch at reduced cost. Students who qualified for this are considered of low SES, those not who did not, of high SES. A preponderance of educational research supports the conclusion that high SES students outperform low SES students. However, in my research, no significant statistical differences were found between these groups with regard to engagement or academic achievement.

As concerns the physical environment, I presented the classroom layout on Day 1 (figure 1.2) and the improved version of it on Day 79, the last day of the semester (figure 1.4). To the best of my knowledge, no student was aware that I was making these physical changes to their classroom (e.g., acquiring and painting cabinets, repairing bookshelves, painting walls, etc.).

SOME NOTES FROM CHAPTER 2: GOING NATIVE: MY FIRST TEN DAYS IN CHEMISTRY

In those very first days in Mrs. Waters' Block 1 Chemistry class, my major concern was whether or not I would be accepted as a fellow student. Failure to gain acceptance within a few days could invalidate the research or at least diminish its validity to some degree. Furthermore, I didn't know how I was likely to affect Mrs. Waters. I did not know her well enough to have earned her trust, nor had I ever observed her teaching prior to that first day. Would she be changing her curriculum and pedagogical strategies in an effort to portray a more structured and rigorous environment given my presence in her classroom? Could I handle her assignments? Would she find my work sub-standard and make my low performance a topic in the teachers' lunchroom or on Facebook? As a participant observer, how could I (or should I) make suggestions to her in real-time without appearing to undermine her authority or expertise?

In the first ten days, I found:

- Mrs. Waters' expectations were consistent as detailed in her syllabus and explained on the first day of class.
- Parental involvement was initiated early on. Parents were required to sign her course syllabus and thereby acknowledge that they had received and read it.
- Students filled out a questionnaire that asked for their email address, and information on their hobbies and interests.
- The class agenda, including the various activities and learning objectives were posted on the whiteboard each day. Students were frequently reminded of standards specific to their lessons as these were referenced in these resources.
- Bell-to-bell teaching happened regularly. Little or no time was lost in transition to new tasks.
- The lessons were engaging and there were ample with hands-on learning opportunities.
- There was a "Problem of the Day" and/or "Chemical of the Day."
- Homework was assigned each night and checked the following day.
- Mrs. Waters assigned membership in groups and varied it with each activity. No students complained about their assignments to groups.
- Wednesdays were established as *Study Island* days; Thursdays were lab days, and quizzes were administered on Fridays. Mrs. Waters had clearly built a consistent structure into the week.
- A pre-test was administered early in the semester to establish a baseline for student knowledge of chemistry.
- Mrs. Waters rarely sat at her desk during class—only when we were taking a quiz or a test. With the changes I made to the physical environment (e.g., acquisition of storage cabinets, removal of obstructing furniture, etc.), she was able move around the entire room and have a full, unobstructed view of each student.

- On Day 9, based on a suggestion from my research, Mrs. Waters initiated the first flexible homework assignment.
- All of the students appeared to take the class very seriously. At the end of the first 10 days, I noted in my journal: "Mrs. Waters has already gained the respect of her students as someone who is passionate about her teaching, knowledgeable of content, who provides engaging activities, and cares about successful learning outcomes for them.

SOME NOTES FROM CHAPTER 3: THE AESTHETICS OF LEARNING

Perhaps the most important findings of my 79 days in her class were those related to flexible assignments. Students routinely put in much more effort on these than expected. They more frequently made a connection between their art discipline and these self-chosen assignments. In other words, there was clear and measureable student engagement (e.g., time on task) happening here. They were not only comfortable in presenting their work but also enjoyed sharing their interests and discussing what they had learned with their peers. As I write this, I am recalling Sima's work on the science behind mood rings—something she had always wondered about. And, I also recall Siraj's investigation of the effectiveness of detergents in removing catsup stains.

On at least two occasions, students who were fully prepared to present, asked Mrs. Waters for an extension so as to be able to present on the following day. They wanted to improve upon their work—and their eagerness to do this was prompted by observing the effort and high quality of the work done by their peers. It was clear that students were setting an ever-higher bar for each other. And, all of this active, student-centered learning was made possible by simply letting them have a voice in designing their own homework and project assignments. To be clear, I would not propose that every assignment should be flexible. In fact, I would recommend using these assignments sparingly as was done in Mrs. Waters' class. Doing this can obviate against student burn out. Flexible assignments should be scheduled in such a way that the potentially highly engaging activity is something students look forward to and want to do.

SOME NOTES FROM CHAPTER 4: MRS. WATERS' FORMAL LAB REPORT

One of the tasks that Mrs. Waters required, proved to be most challenging for students, namely, completion of a formal lab report. It had nine required sections. The lab was performed on Day 30 (3/24/11) and the report was due on Day 42 (4/11/11), so there was ample time to complete it. And, she had posted seven resource documents, including work samples, online for students to access 24/7. Also, the computer lab was reserved for several days so that students could work on their reports. Despite all of this support, of the 33 lab reports received across all three of Mrs. Waters' chemistry classes, 45 percent received a grade that was below passing (where passing = 70%). Of the 21 distinct tasks assessed using her Lab Report Scoring System, the mean scores for 14 of them fell below the desired cutoff of 80 percent for proficiency (see table 4.2).

I was able to provide her with a sorted and prioritized list of task scores so that she could address areas of student weakness. The top criteria candidates proved to be the tasks: Additional Data Related, References to Data, Graph Quality, Discussion of Data Quality, and Additional Analysis. On the other hand, there were four tasks on which all of the 33 students earned perfect scores (i.e., 100%): Title Page, Materials, Format, and Writing Mechanics. The mean rubric score for Procedure Completeness was nearly perfect (98%).

A scatter plot revealed a positive, moderate-to-strong, statistically significant correlation between the Lab Report Total Score and Final Chemistry Class Grade. The regression (i.e., predictor) equation turned out to be:

FINAL CHEMISTRY CLASS GRADE = 0.4891 X (LAB REPORT TOTAL SCORE) + 49.4273

Future research would involve further testing the hypothesis that the Lab Report Total Score is a predictor of final chemistry course grades. Factors contributing to this relationship may be the inherent rigor of the course content since it emphasizes the use of math, reading, and writing skills and also favors positive personal work habits such as obedience to classroom rules, compliance with requirements, attention to detail, and perseverance.

If this proves to be a true predictor of outcomes, such a metric could be very useful to Mrs. Waters in monitoring student progress and also in early detection of student needs for remedial work.

As the exit survey revealed, doing lab experiments was a favorite student activity, writing the Formal Lab Report, their least favorite. This creates a challenge for Mrs. Waters—modifying the process to make the writing process more engaging.

My Reflection on Mrs. Waters' Labs

As noted in chapter 4, I participated in all five lab experiments in Mrs. Waters' class and observed several notable consistencies:

- Every lab was preceded by a "Pre-Lab Activity" worksheet. It was given as a homework assignment prior to the day of the lab—following the paradigm of blended learning.
- Safety concerns were always addressed prior to the start of the labs. Safety glasses were provided when the experiment dictated a need for them and their use was mandatory.
- Mrs. Waters handled the most hazardous materials herself.
- Every lab was set up prior to students entering the classroom.
- Mrs. Waters' always assigned lab partners and those assignments varied with each lab.
- Every lab was well choreographed and there was virtually no downtime. For all practical purposes, in all five of them, the "hands-on" activities (i.e., data collection) were completed in the allotted time with very few exceptions. There were times when data analyses, write-ups, etc. were either assigned as homework or expected to be completed during the following class.
- Mrs. Waters was always actively involved in continuously monitoring each lab station to ensure that students were on task, that safety procedures were being followed, and that there were opportunities for inquiry-based conversation with them.
- Every lab that required a significant setup was held on the same weekday—a Thursday. Mrs. Waters had the first period off on that day and this allowed her to set up the room. One exception was the "Electron Probability Lab." It had minimal set-up requirements and so was conducted on a Tuesday.
- Full class discussion of results immediately followed each lab.

SOME NOTES FROM CHAPTER 5: MRS. WATERS' EXIT SURVEY: WHAT DOES THE RESEARCH SAY?

The exit survey that was administered to Mrs. Waters' Block 1 Chemistry class illustrates how a simple survey can elicit valuable information from students regarding their learning experience. Several of the important findings from it were discussed in chapter 5, but it is worth reflecting upon a few of them further.

Mrs. Waters' assignment of reviewing her PowerPoint slides and copying notes off the slides into our notebooks proved to be a bit boring and monotonous. While the organization and content of the slides were excellent, it was not an engaging task (either for myself or for many of the other students). I had learned long ago that simply viewing information (e.g., watching PowerPoint presentations, videos, or movies), copying notes off the board, or listening to lectures does not promote student engagement—it often has the opposite effect of turning them off. However, integrating other activities such as the use of study guides (e.g., targeted questions) together with the PowerPoints, or with movies, or with lectures, can help support student learning. So, early on, I suggested to Mrs. Waters that she include a study guide with her PowerPoints and I provided two model assignments. The students responded well to this and she then created and assigned study guides for all of her PowerPoints throughout the remainder of the semester.

On the exit survey, there was a question that specifically asked students whether or not they preferred to complete a study guide. Further, they were asked whether they felt they learned more by using the study guide or by not using it. As chapter 5 indicates, 68 percent of them preferred to use it, 76 percent felt they learned more buy doing this. Further, as the survey indicated, when students were asked to rank Mrs. Waters' 12 tasks/activities from most to least favorite, simply copying PowerPoint notes off of slides without a study guide was second from the bottom.

The students' responses on the exit questionnaire indicated that their three most favorite activities were (1) doing lab experiments, (2) the Jeopardy game, and (3) Flexible homework/projects. Completing the Formal Lab Report proved to be the least.

A majority of students (more than 60%) expressed an interest in doing science-related assignments in their art classes. Unfortunately, many art teachers have phobias about math and science. But some post-secondary art teacher preparation programs may be responsible for this in that they have either minimal or non-existent math and science coursework requirements.

SOME NOTES FROM CHAPTER 6: THE STATE SCIENCE EXAM

Now that we know the results of Mrs. Waters' students' performance on the state NECAP science exam, (the number of students achieving proficiency or above doubled) it is clear that that her effort paid off. Using NWEA data to determine the areas of student weakness with regards to the three state-assessed core areas (Earth and Space, Life Science, and Physical Science), made it possible for Mrs. Waters to identify the various types of interventions that were necessary at the individual level where preparation for the NECAP science assessments was concerned.

SOME NOTES FROM CHAPTER 7: A TAXONOMY OF TIME

By sitting in Mrs. Waters' class for those 79 days, I was able to observe first-hand all of the minute distractions, interruptions, and inefficiencies that steal away valuable time. I therefore constructed a taxonomy of the different types of time (table 7.2). It was disconcerting to find that, of the 90 instructional days that were theoretically available on the school calendar, in actuality, we had only 64 to work with. Obviously, this is a significant loss to the student. I have carefully detailed the reasons for this lost time and the data provided herein should help Beacon develop a better plan for future semesters.

SOME NOTES FROM CHAPTER 8: MRS. WATERS' BEST CLASSROOM PRACTICES

Every so often, someone offers me tickets to a major sports event (e.g., MLB, NFL, NHL). I go, but spend most of my time at the snack bar where they often have a flat screen TV. My eyesight prevents me from following the motion of the baseball, football, or focusing on the hockey puck from the stadium seats—especially from the free, nose-bleed seats I always end up with! By contrast to this, watching the games on TV provides an up-close view of the action, and I am able to discern in closer detail what is happening.

One of the catalysts to my "Aha!" moment about becoming a participant observer in Mrs. Waters' class was the sudden realization that I had an opportunity to explore Philip Jackson's supposition in *Life in Classrooms*: "A new look at teaching, if there is to be one, seems to require us to move up close to the phenomena of the teacher's world. But such a move, though long overdue, is just the beginning."

Sitting in the front of Mrs. Waters' chemistry class afforded me a clear view of the action on the whiteboard as well as of the interaction between student and teacher. It allowed me to record and study in great detail, the complexity of the teaching and learning process. From this vantage point and throughout my 79 days, I compiled 30 "best practices" of classroom management. In chapter 8, I suggest a timeframe for implementing them.

The data that I collected provided foundational material for Beacon's "Classroom Management: Best Practices." These can now be found in the *Faculty Practices Guide* which is provided to each member of the school's faculty.

SOME NOTES FROM CHAPTER 9: TEACHER OF THE YEAR—IT'S THE CLIMB!

Just about three months after my final day in Mrs. Waters' Block 1 Chemistry class, she was named Beacon Charter High School's Teacher of the Year and nominated for the state teacher of the year. What were the odds that the teacher whose classroom work I had decided to research would possibly come to such a prestigious honor? We had already developed a great professional relationship through the collaborative effort entailed by my research, and were on the way to becoming good friends, so I wasn't surprised when she asked for my advice from time to time. I decided to follow her journey through this process and thus include this adventure as chapter 9 of this work. As an educational researcher, I saw this as a very rare opportunity to capture the process of becoming a state "Teacher of the Year" in real-time, something that has not been addressed in such a manner in the educational research literature to date. Hopefully, the story I tell in chapter 9 conveys the rollercoaster of emotions and obligations that Mrs. Waters experienced with her complex schedule and need to balance the roles of teacher, mother of three children, wife, and nominee for Teacher of the Year. Her willingness to share her thoughts with me (and the reader) on this journey—this climb—made for a much broader and deeper understanding of the demands that were placed upon her both in and out of the classroom. It certainly added perspectives beyond those of an evaluating principal or administrator who merely observes from the back of a classroom with an iPad for one or two days—a process that can only claim a snapshot of the teaching and learning process happening in the classroom.

Recent research by Professor James Flynn, world renowned authority on IQ tests, has found that these scores have risen at a faster rate for women than for men in the last 100 years. In her article "Women do have Higher IQ than Men (Thanks to Multi-Tasking), author Jenni Marsh writes "One theory is that multi-tasking modern women, who juggle careers and raise a family, are becoming more intelligent as a result of their demanding life styles" (Marsh 2012). I'm not sure whether all of this is true, but I know if I were ever asked to state three characteristics that best describe Mrs. Waters, "multi-tasker" would be one of them.

SOME NOTES FROM CHAPTER 10: REFLECTIONS ON THE RESEARCH

As stated in the closing remarks in chapter 10, I never grade reflections but I do carefully consider their content. I was very interested to find out whether my presence in Mrs. Waters' classroom had negatively affected student learning, whether or not I had been accepted as a fellow student, and whether the students felt that they had changed their behavior because of my presence in their class. Their positive comments supported the validity of my research.

With regards to Mrs. Waters' reflection on this, it was interesting to read that she felt that, in the beginning, there was some awkwardness in the situation—as if she were going to have to give "special" lessons because another teacher would be evaluating her virtually every day. This was one of the fears that I hesitated to articulate in the preface of this work. Needless to say, I was pleased that Mrs. Waters quickly realized that I was serious about the research and that it offered some possibilities for gaining insight to student engagement and identifying teacher "best practices." Further, her observation that students accepted me as one of their peers is confirmed by their reflections and also underscores the validity of my research. Finally, Mrs. Waters' expressed appreciation for the positive learning experiences that derived from use of data to drive instruction (e.g., analysis using the Formal Lab Report Scoring System, Student Exit Survey) and from some of my suggestions for tweaking of assignments (e.g., creating study guides aligned with PowerPoint presentations, designing a graphic organizer template for Chemical of the Day), from realizing how valuable the flexible assignments proved to be in so far as they gave students a voice in creating their own homework and projects. All of these things can play a role in deepening student engagement in learning and students' art interests can as well. In sum, participant observation can positively impact upon the culture of a classroom and upon students' success in learning.

This study yielded a wealth of data, including many examples of student engagement and an impressive list of Mrs. Waters' best teaching practices.

Key findings

- Students know more than they say.
- Students will do more when they are engaged.
- Aesthetics plays a major role in student engagement.
- Increased student engagement was observed with flexible homework and flexible project assignments.
- The hands-on activities: lab experiments and flexible assignments, were amongst the top three favorite student activities.
- Students generally don't complain about the physical environment of their classroom–they adjust to it.
- Nobody complained about working with others when Mrs. Waters routinely assigned membership in group work.

- Once a daily routine is established and becomes an expectation, most students don't need to be reminded of it. One case in point is the expectation that upon entering class, they check the whiteboard for the class agenda and begin work on the Problem of the Day and/or Chemical of the Day.
- Several Students took advantage of many of the extra credit opportunities that Mrs. Waters offered. This was another indicator of their engagement.
- Teacher collaboration, such as between Mrs. Waters and myself, builds social capital which in turn improves human capital (i.e., individual) with the overall outcome of increasing professional capital.
- The charter school environment, especially a non-union one, can provide opportunities for educational research such as I have described here within.

A FEW COMMENTS ON PARTICIPANT OBSERVATION RESEARCH

In the preface, I mentioned that there were a few key resources to which I could refer to as I immersed myself in this research project that drew upon relatively rare methodology. My initial concern was whether or not participant observation was an appropriate methodology for exploring student engagement and teacher best practices. To address it, I referred to Danny L. Jorgensen's excellent handbook entitled *Participant Observation: A Methodology for Human Studies* (Sage).

He provides six minimal conditions for appropriate use of the participant observation methodology: (1) the research question involves human meanings and interactions from the insiders' perspective; (2) the phenomenon being studied is observable within an everyday life situation or setting; (3) the researcher has access to an appropriate setting; (4) the phenomenon is sufficient in size and location to be treated as a case; (5) the research questions are appropriate for case study; and, (6) the research question can be addressed by qualitative data gathered by direct observation and other means.

Certainly, my role as a student who was studying other students' engagement and their interactions with their teacher would satisfy the first requirement. Attending chemistry class everyday would be the best opportunity to capture this phenomenon (2). Since I am a teacher at Beacon Charter High School, my schedule would allow me to attend class every day (3). The fact that Mrs. Waters' Chemistry Block 1 class routinely met in Room 205 with the same participants qualifies as a case (4). My research questions relative to student engagement and teacher best practices can be addressed in a case study (5). And, finally, inherent to my participating in and observing the classroom interaction would be the multiple opportunities to collect qualitative data (6).

As Jorgensen states, "Ultimately, the methodology of participant observation aims to generate practical and theoretical truths about human life grounded in the realities of daily existence" (p. 14).

A second concern revolved around the ethics of my research: Was I breaking any rules? Educational research does not normally involve putting participants at risk of physical harm, nor would such research ever be approved. My greater concern was that I did not want to negatively affect the mental state of the students or in any way interfere in their education. After all, my larger goal was to understand the learning process in a classroom setting and explore ways of refining and enhancing it. Jorgensen does address issues of ethics in his handbook. He distinguishes between research in experimental psychology and medicine that involves human subjects and "The participant observer interacts with people under the ordinary conditions of their daily lives much like any other participant. . . . Consequently, the participant observer has no more or less of an ethical obligation to the people encountered in the course of research than she or he would have under other everyday life circumstances" (p. 28). And, he further states that "the researcher is not necessarily obligated to inform people of research intentions, or even protect them from possible harmful consequences." However, ethical codes generally respect the right to privacy, confidentiality, and freedom from exploitation.

With regards to my concern about the possibility that I would negatively affect teaching and learning in Mrs. Waters' class, her reflections and those of the students (ch. 10), clearly support the fact that my presence was conducive to a positive experience.

In writing this book, I have been particularly aware of the need to honor the rights to privacy, confidentiality, and freedom from exploitation for everyone involved in my research. Only student first names are used throughout and pseudonyms were chosen where circumstances necessitated this. Prior to writing, I had asked every student to sign a consent form which gave me permission to use their first names to indicate their dedication and share and properly attribute their work to them. Where the adults were concerned, again in potentially sensitive situations I used a pseudonym. And of course, I cannot thank Mrs. Waters' and her family enough. Her co-operation allows for a closer look at a teacher's life outside of the classroom, something that is rarely discussed in the research literature.

Finally, Jorgensen discusses various ways a participant observer may depart from the field of study. Such a departure can be either a great relief if the experience was unpleasant or a sad time if friendships had been established and people will be missed. In my case, the only element of sadness was the fact that the 79 consecutive days with a wonderful class of students had to come to an end. Now, they have all graduated from Beacon and most are in college. I hear from them occasionally—we always enjoy some good conversation about our shared experiences in Mrs. Waters' Chemistry Block 1 class.

Since completing this research project, I have teamed up with her on several occasions to present our collaborative findings at various conferences. Of course, she is the Rhode Island 2013 Teacher of the Year—I'm there to set up the projector!

WHAT NEXT?

The short scholarly answer to this question is that the findings from my research—on student engagement, teacher best practices, data-driven instruction, student-centered and inquiry-based teaching and learning, and participant observation research—now need to be disseminated. Flexible assignments could be tried in other content areas. And, the best practices observed in Mrs. Waters' class can be shared. Further, the power of collaboration between teachers (and administrators) with regards to improving the profession and students experiences cannot be emphasized enough.

NOTE

1. Alexis Carrel (1873–1944), French surgeon and biologist, was awarded the Nobel Prize in Physiology or Medicine in 1912. Cited in *The Harper Book of Quotations: Third Edition* edited by Robert I. Fitzhenry (1993, 402).

Appendix

A Prime Connection?

A discovery is said to be an accident meeting a prepared mind.

—Albert Szent- Györgyi[1]

ABOUT THIS WORK

The following discussion of electron configurations resulted from a homework assignment (mandatory-flexible type) made by Mrs. Waters to her Chemistry Block 1 class. This topic is mentioned here because I was initially confused by a handout she had distributed to our class prior to a lecture on writing electron configurations (figure A.1). It is a commonly used diagram used to introduce electron configurations based on the Madelung Rule (named after Erwin Madelung) in a high school chemistry curriculum for predicting the order in which electrons fill the orbitals of atoms in the *Quantum Mechanical Model*[2] following the *Pauli exclusion principle*[3]; *Aufbau principle*[4]; and, *Hund's rule*[5]. During my first exposure to this diagram, I found the process both time consuming and such that one could easily make mistakes in writing down the electron configurations. Further, as this figure indicates, the handout provided no direct way of quickly identifying the electron configuration for a given atomic number (i.e., an element from the Periodic table). I therefore immediately started working on a more efficient graphic, a shortcut that would entail a process less prone to causing errors.

Remember to start at the beginning of each arrow, and then follow it all of the way to the end, filling in the sublevels that it passes through. In other words, the order for filling in the sublevels becomes; 1s, 2s, 2p, 3s, 3p, 4s, 3d, 4p, 5s, 4d, 5p, 6s, 4f, 5d, 6p, 7s, 5f, 6d,7p.

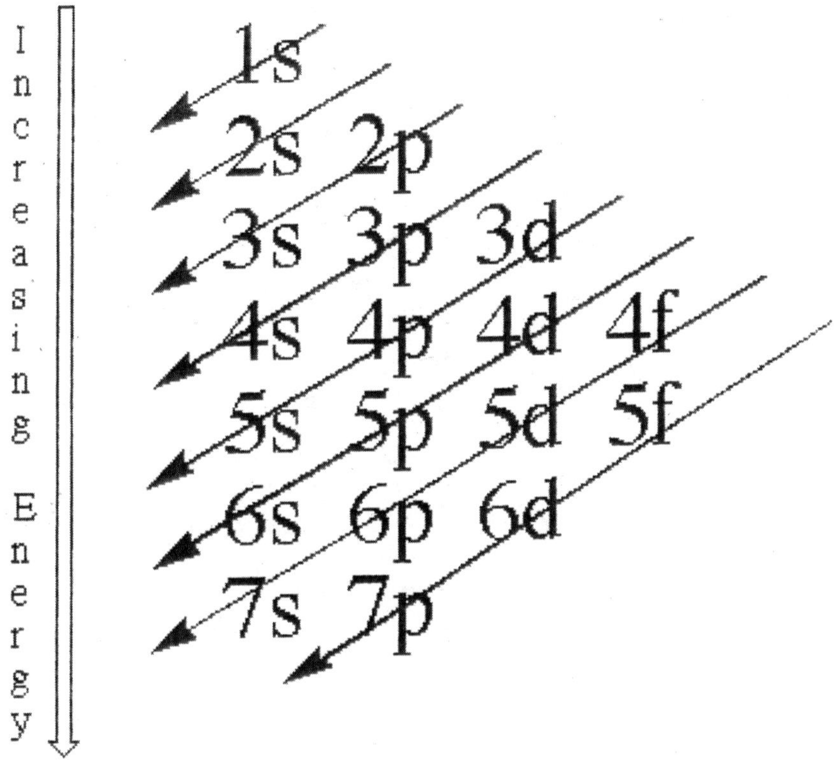

Figure A.1. A traditional method for teaching the ordering of electron configurations.

BACKGROUND INFORMATION

The distribution of electrons in the atom is commonly described by a finite, ordered linear array. It consists of an ordered list of one or more non-repeating combinations which are comprised of a positive integer coefficient, a lower case letter, and a positive integer superscript. Each combination is written in the form:

[positive integer coefficient][a lowercase letter]$^{\text{[a positive integer superscript]}}$ where,

- the positive integer coefficient represents the *energy level.*
- the lowercase letter designates the *subshell.* The subshells are named s, p, d, and f.
- the positive integer superscript is the *number of electrons in the subshell.*

This linear array is referred to as the *electron configuration.*

Example 1. The electron configuration for the element neon (Ne) with its 10 electrons is:

$$1s^2 2s^2 2p^6$$

Example 2. The electron configuration for the element magnesium (Mg) with its 12 electrons is:

$$1s^2 2s^2 2p^6 3s^2$$

Magnesium has the atomic number 12, i.e. it has 12 electrons (and 12 protons). Since no two elements have exactly the same atomic number (i.e., the same number of electrons), each element has its own unique electron configuration. The "1s," "2s," "2p," and "3s" in the electron configuration for magnesium are called orbital names. The respective superscripts attached to these orbital names designate the number of electrons in the *orbital* (to be defined later). Although the superscripts appear as exponents, they are not reflective of a mathematical operation. The sum of all of the integer superscripts in the electron configuration is equal to the atomic number (i.e., to the total number of electrons) of a specific element.

The electron configuration is based on the order in which the electrons fill the orbitals in the Quantum Mechanical Model. Although the term orbital sounds like orbit, it represents something quite different. An orbit refers to an object's path in a more or less circular direction about some center point such as the path that the earth follows around the sun. The position of the Earth relative to the Sun is known at any point as is its momentum. But, this is not the case for electrons. According to the *Heisenberg Uncertainty Principle*, it is not possible to be absolutely certain simultaneously of both the position and momentum of an electron. However, we do know the region within which it will be found—a shape referred to as an *orbital*—and they have several different shapes.

The rows in figure A.1 represent different energy levels, i.e. energy moves downward from the top row to the bottom one. There are seven energy levels. The columns represent the subshells ordered from left-to-right as s, p, d, and f. Arrows refer to the order in which the electron configuration is written following the Madelung Rule. On the basis of this diagram, the following order of orbital names results:

1s, 2s, 2p, 3s, 3p, 4s, 3d, 4p, 5s, 4d, 5p, 6s, 4f, 5d, 6p, 7s, 5f, 6d, 7p

WRITING THE ELECTRON CONFIGURATION USING THE MADELUNG DIAGRAM

To use the diagram in figure A.1, start at the beginning of the top arrow and write any orbital names passed as one proceeds downward. Jump to the beginning of the next lower arrow and again write down orbital names. However, there is an assumption behind working with the particular diagram in figure A.1, namely, that "s" orbitals are known to hold two electrons, "p" orbitals, six, "d" orbitals, 10 , and "f" orbitals, 14. So, as in writing the electron configurations for a known atomic number, one must add the electrons in each orbital in figure A.1 until one reaches the desired total. But, all of this is somewhat inconvenient. Consider magnesium with its atomic number of 12:

From the top,
Write: $1s^2$ (sum of electrons = 2);
after passing down the next lower arrow,
Write: $1s^22s^2$ (sum of electrons = 4);
finally, after passing down the next lower arrow,
Write: $1s^22s^22p^63s^2$ (sum of electrons = 12)

I have now accounted for all 12 electrons and have written down the electron configuration for magnesium. However, as it turns out, the diagram is very accurate for predicting the electron configuration as compared with actual experimental results. Indeed, it produces only two minor discrepancies in the first 40 elements: chromium (Cr) and copper (Cu).

Chromium (24 electrons):
predicted electron configuration: $1s^22s^22p^63s^23p^64s^23d^4$
actual electron configuration: $1s^22s^22p^63s^23p^64s^13d^5$

Copper (31 electrons):
predicted electron configuration: $1s^22s^22p^63s^23p^64s^23d^9$
actual electron configuration: $1s^22s^22p^63s^23p^64s^13d^{10}$

The frequencies of discrepancies between what is predicted and what experiment reveals increase with the atomic number. However, the diagram in figure A.1 is still quite a good predictor of atomic numbers greater than 40.

Handling an atomic number of 12 works fairly well with this method, but how about radium (Ra) with its atomic number of 88?

Using figure A.1 and following the arrows until I have accounted for all 88 electrons, I write:

$$1s^2 2s^2 2p^6 3s^2 3p^6 4s^2 3d^{10} 4p^6 5s^2 4d^{10} 5p^6 6s^2 4f^{14} 5d^{10} 6p^6 7s^2$$

Now, adding the superscripts as a check:

$$2+2+6+2+6+2+10+6+2+10+6+2+14+10+6+2 = 88$$

As the atomic number increases by one unit, each electron configuration builds upon the preceding one. Given this fact, there is an identifiable shortcut that can sometimes be used in writing an electron configuration. It involves replacing the first portion of the long string by the electron configuration of a noble gas (e.g., neon, argon, krypton, and xenon).

Example 1: The standard electron configuration for Magnesium (Mg), with atomic number 12, is written as:

$$1s^2 2s^2 2p^6 3s^2.$$

Since, noble gas neon (Ne), with atomic number 10, has the electron configuration:

$1s^2 2s^2 2p^6$

By substitution, we can write the shorthand version as:

$[Ne]3s^2$

Example 2: the standard electron configuration for Radium (Ra), with atomic number 88, is written as:

$$1s^2 2s^2 2p^6 3s^2 3p^6 4s^2 3d^{10} 4p^6 5s^2 4d^{10} 5p^6 6s^2 4f^{14} 5d^{10} 6p^6 7s^2$$

Since, noble gas radon (Rn), with atomic number 86, has the electron configuration:

$$1s^2 2s^2 2p^6 3s^2 3p^6 4s^2 3d^{10} 4p^6 5s^2 4d^{10} 5p^6 6s^2 4f^{14} 5d^{10} 6p^6$$

We can write the shorthand version as:

$$[Rn]7s^2$$

However, although some convenience is gained in terms of writing the electron configuration and the chance of making mistakes is lower, this abbreviated method does not efficiently link the electronic configuration with an atomic number, nor vice-versa. Again, for atomic numbers of greater than 40, some predicted electron configurations vary from the experimentally observable ones.

A PROPOSED NEW METHOD FOR WRITING PREDICTED ELECTRON CONFIGURATIONS FOR THE ELEMENTS (THIS IS RECOMMENDED ONLY AFTER STUDENTS HAVE LEARNED FUNDAMENTAL CONCEPTS)

A new method for deriving and writing predicted electron configurations for the first 118 elements is presented here. It allows one to determine the predicted notation in a matter of seconds and was developed by expanding the diagram in figure A.1, i.e. by following the energy directions in order of increasing atomic number. For the presentation of my flexible homework project in Mrs. Waters' Chemistry Block 1 class, I created a poster-sized diagram of my proposal. This made it more readable. A portion of this diagram is shown in figure A.2. It generates the first 58 electron configurations. The actual poster-sized diagram that was submitted to Mrs. Waters generated 118 predicted electron configurations.

Procedure (refer to the key in figure A.2; figure A.3 provides an enlarged view)

Step 1. From figure A.2, find the row that contains the square box with the atomic number of interest. Follow the direction of the arrows in the row from right to left and note that the atomic numbers in the square boxes increase by exactly one. The number in the circle above the square box is the number of electrons that must be written in the open circle in superscript position to the left of the atomic number being derived. It represents the number of electrons in the outer shell. The number above the open circle is the maximum number that can be placed in the open circle as it represents the maximum number of electrons in the particular orbital. This is the last orbital name that will appear in the desired electron configuration.

Step 2. Add this orbital name to the electron configuration that is written out at the right end of the previous row to complete the predicted electron configuration.

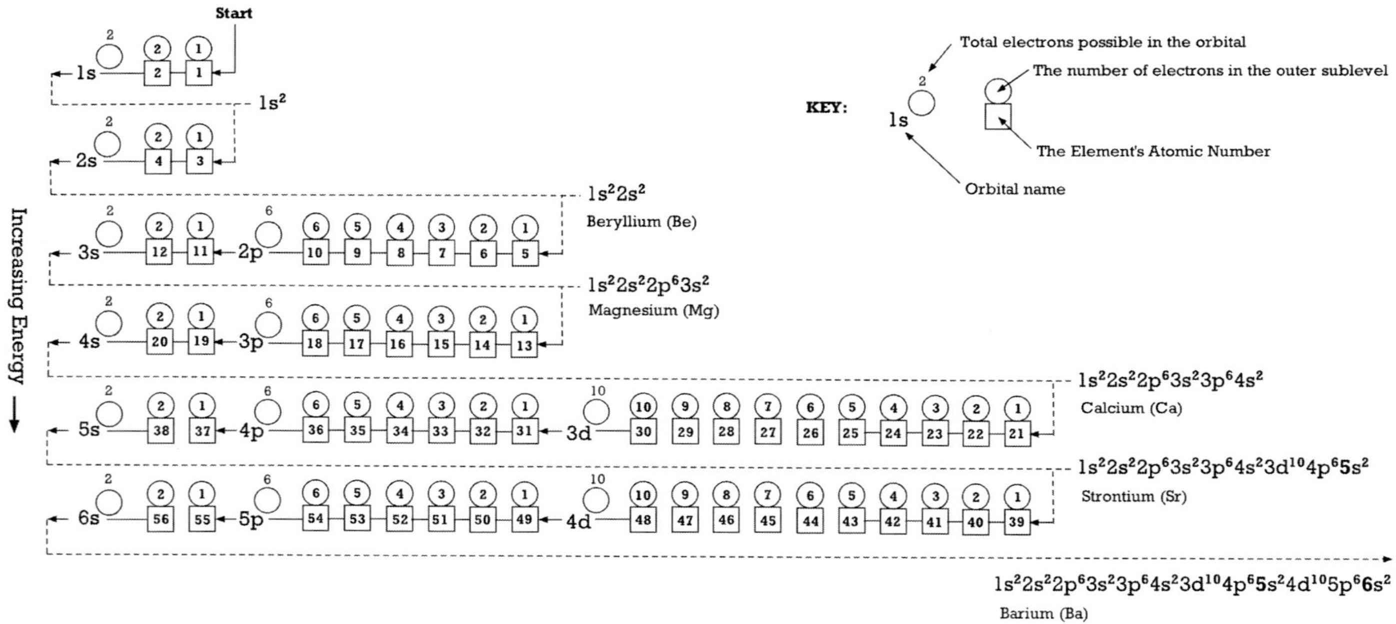

Figure A.2. A new method for writing predicted electron configurations.

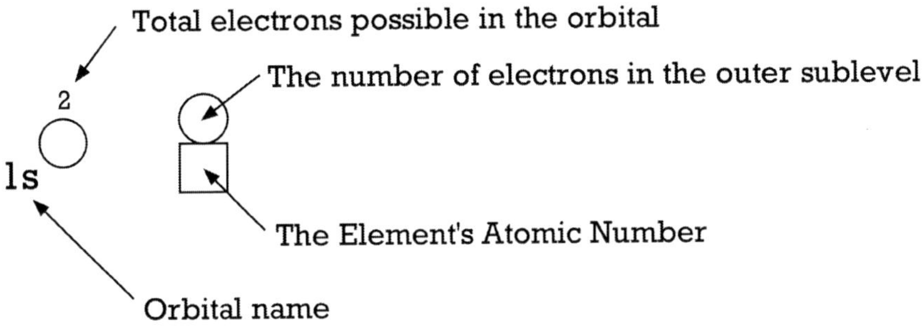

Figure A.3. Enlarged Key.

For example, to write the electron configuration for the element carbon, with an atomic number 6 (i.e., 6 electrons):

Step 1. Find the third row with the square box that has a 6 in it. Next, note that there is a number 2 in the circle above it. Place the 2 in the open circle to the left. This will result in the orbital name $2p^2$.

Step 2. Add this orbital name with the superscript to the previous "end-of-row" string (i.e., $1s^22s^2$) to obtain the electron configuration for carbon as $1s^22s^22p^2$.

In the case where there are open circle(s) in the same row and to the right of the square box with the desired atomic number, the open circle(s) to the right are filled with the maximum allowed number of electrons noted above the open circle superscript.

It is feasible to apply a shortcut method such as the one building upon the noble gases as previously described. However, as noted in figure A.2, instead of the noble gas electron configurations, Group IIA element electron configurations (e.g., beryllium, magnesium, calcium, strontium, barium, radium) can be used. These are listed at the right end of the rows in figure A.2.

AN "AHA!" MOMENT: A POSSIBLE DISCOVERY?

As I was laying out the diagram for my homework presentation (figure A.2) there seemed to be an observable pattern between atomic numbers that were *prime* (i.e., numbers divisible by only "1" and itself) and a superscript of "1" in the last orbital name in the electronic configuration, which indicates only one electron in the outer subshell. Atomic numbers are shaded in figure A.4 because they are primes. Of the first 56 elements shown in figure A.4, 47% (8/17) of the possible prime numbered elements have one electron in their outer subshell.

Below is the pattern of prime atomic numbers with predicted electron configurations that contain the last orbital names with a superscript equal to "1." It covers all of the 118 elements I considered:

Atomic Number = 1: $1s^1$
Atomic Number = 5: $1s^2 2s^1$
Atomic Number = 11: $1s^2 2s^2 2p^1$
Atomic Number = 13: $1s^2 2s^2 2p^6 3s^2 3p^1$
Atomic Number = 19: $1s^2 2s^2 2p^6 3s^2 3p^6 4s^1$
Atomic Number = 31: $1s^2 2s^2 2p^6 3s^2 3p^6 4s^2 3d^{10} 4p^1$
Atomic Number = 37: $1s^2 2s^2 2p^6 3s^2 3p^6 4s^2 3d^{10} 4p^6 5s^1$
Atomic Number = 71: $1s^2 2s^2 2p^6 3s^2 3p^6 4s^2 3d^{10} 4p^6 5s^2 4d^{10} 5p^6 6s^2 4f^{14} 5d^1$
Atomic Number = 89: $5d^{10} 6p^6 7s^2 5f^1$
Atomic Number = 103: $5f^{14} 6d^1$

Even though the number "1" is not considered by mathematicians to be a prime number, (this makes the Unique Prime Factorization Theorem[6] hold) I considered the atomic number "1" to be a prime, acknowledging that nature probably doesn't honor such man-made constraints.

As noted, I considered 118 elements. This observation did not hold true for every prime atomic number but there was definitely a statistically disproportionate, greater number of prime atomic numbers which had only one electron in their outer subshell.

After I presented my homework assignment to the class, Mrs. Waters hung my poster on the wall. In the days that followed, several students chose to use this method in preference to the traditional diagram. They stated that they found it much more efficient.

So, discovering what seems to be a connection between prime numbers and the Quantum Mechanical Model was an "Aha!" moment for me.

It is worth noting that, for over 2000 years, the distribution of primes in our complex number system has remained one of the greatest unsolved mysteries in mathematics. It has been demonstrated by mathematical proof that there is no single equation that generates and predicts all primes. Yet mathematicians feel strongly that there has to be some master plan for the order and frequency in which they appear. Maybe there is one that pertains between prime numbers and the Quantum Mechanical Model?

As it turns out, however, my "Aha!" moment was literally a "moment." I conducted a very thorough search of the literature and things were looking good for a while. But, just before I was about to call a press conference, I came across an obscure reference to a possible connection between atomic theory and prime numbers. In "On the Limit for the Periodic Table of the

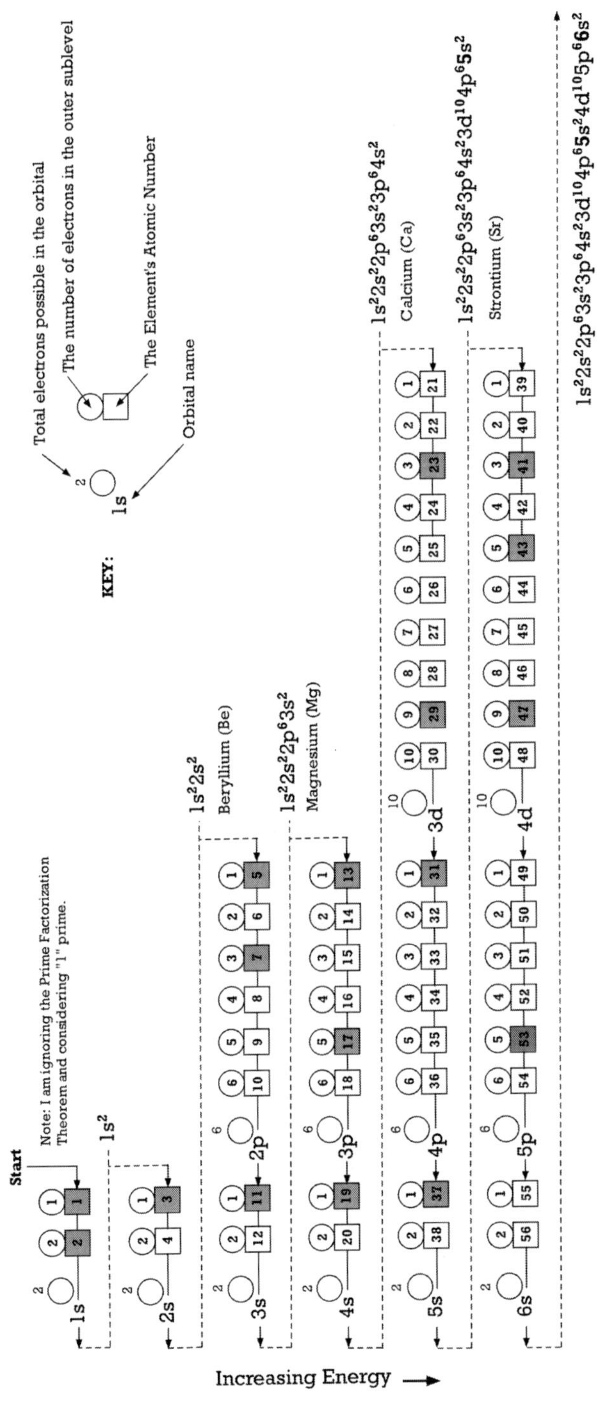

Figure A.4. Location of prime atomic numbers with one electron in the outer subshell.

Elements" by Jiang Chun-xuan of The People's Republic of China, the author discusses a link between the stability of an element and prime numbers (Chun-xuan 1998).

This notwithstanding, the important lesson I took away from Mrs. Waters' "flexible" assignment was the realization that I had become very engaged in learning, that I had dedicated many hours to it. And, I suspect that my experience paralleled that of my classmates—of Brittany when she chose to study the behavior of yeast in the baking of pizza; of Sarah who baked cupcakes and frosted them with symbols of the elements from the periodic table; and of Robert, whose poster-sized painting depicted the life cycle of a star.

A FINAL REMARK ON THE HOMEWORK ASSIGNMENT ON ELECTRON CONFIGURATIONS

For any nuclear scientists out there who may be reading this work and profusely cursing the fact that I am completely off track with my reasoning here, please excuse the wanderings of of my adolescent mind. After, all, I was just a high school chemistry student in Mrs. Waters' Chemistry Block 1 class trying to eke out a passing grade!

NOTES

1. Albert Szent- Györgyi (1893–1986), Hungarian physiologist who was awarded the Nobel Prize in "Physiology or Medicine" in 1937. Cited in Andrew G. Szent-Györgyi's *The Early History of the Biochemistry of Muscle Contraction* published June 1, 2004 // JGP vol. 123 no. 6, 631-641. The Rockefeller University Press, DOI: 10.1085/jgp.200409091. Copyright © 2004, by The Rockefeller University Press.

2. The Quantum Mechanical Model is accepted as the most current one for understanding of the atom. It is based on Neil Bohr's 1913 planetary model of the atom, on which the orbit of an electron is defined in terms of spherical orbits referred to as *shells* or *energy levels* about a small dense nucleus which contain the neutrons and protons. Building upon Bohr's proposal, Werner Heisenberg offered his uncertainty principle in the 1920's. It states that it is only possible to calculate the probability that an electron will be found within a given space and suggested that they travel in diffuse clouds around the nucleus of an atom. In 1926, Erwin Schrödinger, in treating electrons as waves, developed a mathematical model for their distribution within an atom. The Schrödinger model defines the probability that an electron can be found in a given region of space at a given time.

3. The Pauli exclusion principle, named after Austrian chemist Wolfgang Pauli who, in 1925, discovered that no two electrons in an atom can have the same set of

four quantum numbers. The consequence of this is that no atomic orbital can contain more than two electrons.

4. The Aufbau principle. The German word *Aufbau* means "building up." It states that the electrons fill orbitals with the lowest levels of energy first (Myers, Oldham, and Tocci 2004).

5. Hund's rule. Named after German physicist Friedrich Hermann Hund (1896–1997)—a major contributor to quantum theory—this principle states that every orbital in a subshell is singly occupied with one electron before any one orbital is doubly occupied, and that all electrons in singly occupied orbitals have the same spin.

6. Unique Prime Factorization Theorem. The theorem states that every integer greater than "1" is either itself prime, or a unique product of primes. For example the composite (i.e., nonprime) number 12 is the product of 2 x 2 x 3 or 2^2 x 3. There is no other product of primes that will produce 12.

Bibliography

BOOKS AND JOURNAL ARTICLES

Anderson, Lorin. W. 1985. "Time and Timing." In *Perspectives on Instructional Time.* Charles W. Fisher, and David C. Berliner, Eds. 157–168. New York: Longman.

Backes, Charles E., and Iris C. Ellis. 2003. "The Secrets of Classroom Management." *Techniques* 78(5): 22–25.

Bambrick-Santoyo, Paul. 2010. *Driven by Data: A Practical Guide to Improve Instruction.* San Francisco: Jossy-Bass.

Becker, Howard S., and Blanche Geer. 1969. "Participant Observation and Interviewing: A Comparison. In *Issues in Participant Observation: A Text and Reader.* George J. McCall, and J. L. Simmons, Eds. 322–331. Reading, Massachusetts: Addison-Wesley.

Berliner, David C. 1990. "What's all the Fuss About Instructional Time?" In *The Nature of Time in Schools: Theoretical Concepts, Practitioner Perceptions.* Mariam Ben-Peretz and Rainer Bromme, Eds. 3–35. New York: Teachers College Press.

Beyer, William A., Jan Mycielski, and Gian-Carlo Rota. eds. 1974. *Stanislaw Ulam: Sets, Numbers, and Universes: Selected Works.* Cambridge: MIT Press.

Birk, James P., Larry McGrath, and S. Kay Gunter. 1981. "A General Chemistry Experiment for the Determination of the Oxygen Content of Air." *Journal of Chemical Education*, 58(10): 804-805.

Birkoff, George David. 1956. "Mathematics of Aesthetics" In *The World of Mathematics.* vol. 4. James R. Newman, Ed. New York: Simon and Schuster.

Block, John H. 1980. "Success Rate." In *Time to Learn.* Carolyn Denham and Ann Lieberman, Eds. 95-106. Sacramento, CA: California Commission for Teacher Preparation and Licensing.

Borg, Walter. R. 1980. "Time and School Learning." In *Time to Learn.* Carolyn Denham and Ann Lieberman, Eds. 33–72. Sacramento, CA: California Commission for Teacher Preparation and Licensing.

Bowman, Richard. 1983. "Effective Classroom Management: A Primer for Practicing Professionals." *Clearing House* 57(3): 116–118.

Brookhart, Susan M. 2013. *How to Create and Use Rubrics: for Formative Assessment and Grading*. Alexandria, VA: ASCD.

Butler, John D. 2011d. *Faculty Practices Guide 2011-2012*. Woonsocket, RI: Beacon Charter High School.

Caldwell, Janet H., William G. Huitt, and Anna O. Graeber. 1982. "Time Spent in Learning: Implications from Research." *The Elementary School Journal*, 82(5): 470-480.

Carroll, John B. 1985. "A Model of School Learning." In *Perspectives on Instructional Time*. Charles W. Fisher, and David. C. Berliner, Eds. 59–72. New York: Longman.

Clark, Robert L., Michael P. Clough, and Craig A. Berg. 2000. "Modifying Cookbook Labs: A Different Way of Teaching a Standard Laboratory Engaging Students and Promotes Understanding." *The Science Teacher*, 67(7): 40–43.

Cooper, Harris. 1989a. *Homework*. White Plains, NY: Longman.

———. 1989b. "Synthesis of research on homework." *Educational Leadership*, 47(3): 85-91.

Cooper, Harris, Jorgianne Civey Robinson, and Erika A. Patall. 2006. "Does homework improve academic achievement? A synthesis of research, 1987-2003." *Review of Educational Research*, 76(1): 1-62.

Danielson, Charlotte. 1996. *Enhancing Professional Practice: A Framework for Teaching*. Alexandria, VA: ASCD (Association for Supervision and Curriculum Design).

Deters, Kelly Morgan. 2003. "What Should We Teach in High School Chemistry." *Journal of Chemical Education* 80(10): 1153–1155.

Eisenhart, Margaret, and Hilda Borko. 1993. *Designing Classroom Research: Themes, Issues, and Struggles*. Boston: Allyn and Bacon.

Emmer, Edmund T., and Carolyn M. Evertson. 1981. "Synthesis of Research on Classroom Management." *Educational Leadership* 38(4): 342–347.

Emmer, Edmund T., Carolyn M. Evertson, and Linda M. Anderson. 1980. "Effective Classroom Management at the Beginning of the School Year." *The Elementary School Journal* 80(5): 219–231.

Evertson, Carolyn M., and Edmund T. Emmer. 1982. "Effective Classroom Management at the Beginning of the School Year in Junior High Classes." *Journal of Educational Psychology* 74(4): 485–498.

Fischel, William A. 2009. *Making the Grade: The Economic Evolution of American School Districts*. Chicago: University of Chicago Press.

Flannery, Maura. 1992. "Using Science's Aesthetic Dimension in Teaching Science." *Journal of Aesthetic Education* 26(1): 1–15.

Friedland, Aaron. J. 1970. *Puzzles in Math & Logic*. New York: Dover.

Gill, Brian P., and Steven L. Schlossman. 2000. "The Lost Case of Homework Reform." *American Journal of Education* 109(1): 27-62.

———. 2004. "Villain or Savior? The American Discourse on Homework, 1850-2003." *Theory into Practice* 43(3): 174-181.

Greene, Maxine. 2001. *Variations on a Blue Guitar: The Lincoln Center Institute Lectures on Aesthetic Education*. New York: Teachers College Press.

Hargreaves, Andy, and Michael Fullan. (2012). *Professional Capital: Transforming Teaching in Every School*. New York: Teachers College.

Harlacher, Jason E., Nelson Walker, Nancy J., and Amanda K. Sanford. 2010. "The 'I' in RTI: Research-Based Factors for Intensifying Instruction." *Teaching Exceptional Children* 42(6): 30–38.

Harnischfeger, Annegret, and David E. Wiley. 1985. "Origins of Active Learning Time." In *Perspectives on Instructional Time*. Charles W. Fisher, and David C. Berliner, Eds. 133–156. New York: Longman.

Jackson, Philip. W. (1968) 1990. *Life in Classrooms*. New York: Holt, Rinehart and Winston. Reprint, New York: Teachers College Press. Citations refer to the Teachers College Press edition.

Jorgensen, Danny. L. 1989. *Participant Observation: A Methodology for Human Studies*. Vol. 15 of *Applied Social Research Methods Series*. Newbury Park, CA: Sage.

King, Jerry P. 1992. *The Art of Mathematics*. Mineola, NY: Dover.

Kohn, Alfie. 2006. The Homework Myth: Why Kids get too much of a Bad Thing." Philadelphia, PA: Da Capo Press.

Lemov, Doug. 2010. *Teach Like a Champion: 49 Techniques that Put Students on the Path to College*. San Francisco: Jossey-Bass.

London, Robert. 1989. *Nonroutine Problems: Doing Mathematics*. Providence, RI: Janson Publications.

Luntz, Frank I. 2007. *Words That Work: It's Not What You Say, It's What People Hear*. New York: Hyperion.

Martin, Michael O., Ina V. S. Mullis, Pierre Foy, and Gabrielle M. Stanco. 2012. "TIMSS 2011 International Results in Science." International Association for Evaluation of Educational Achievement (IEA). Chestnut Hill, MA: TIMSS & PIRLS International Study Center.

Marzano, Robert J. 2003. *What Works in Schools: Translating Research into Action*. Alexandria, VA: ASCD (Association for Supervision and Curriculum Development).

Marzano, Robert J., and Jana S. Marzano. 2003. "The Key to Classroom Management." *Educational Leadership* 61(1): 6–13.

McCall, George. J. and Simmons, J. L. eds. 1969. *Issues in Participant Observation: A Text and Reader*. Reading, MA: Addison-Wesley.

McManus, I. C., D. Edmondson, and J. Rodger. 1985. "Balance in Pictures." *British Journal of Psychology* 76(1): 311–324.

Mulroy, David. 2003. *The War Against Grammar*. Portsmouth, NH: Boynton/Cook.

Munro, Thomas. 1956. *Toward Science in Aesthetics*. Indianapolis, IN: Bobbs-Merrill.

Myers, R. Thomas, Keith B. Oldham, and Salvatore Tocci. 2004. *Chemistry*. New York: Holt, Rinehart, and Winston.

Olson, John. 1990. "The Influence of Time in the Science Curriculum." In *The Nature of Time in Schools: Theoretical Concepts, Practitioner Perceptions.* Miriam Ben-Peretz, and Rainer Bromme, Eds. 227-239. New York: Teachers College Press.

Overmier, J. Bruce, and Martin E. P. Seligman. 1967. "Effects of Inescapable Shock Upon Subsequent Escape and Avoidance Responding." *Journal of Comparative and Physiological Psychology* 63(1): 28–33.

Pavlov, Ivan. P. 1960. *Conditioned Reflexes: An Introduction of the Physiological Activity of the Cerebral Cortex.* New York: Dover (Original work published 1927).

Pedota, Paul J. 2007. "Strategies for Effective Classroom Management in the Secondary Setting." *Clearing House* 80(4): 163–166.

Peterson, Christopher, Steven F. Maier, and Martin E. P. Seligman. 1993. *Learned Helplessness: A Theory for the Age of Personal Control.* New York: Oxford University Press.

Pilkington, Robert, and Michael Skeldon. 2006. *Beacon Charter High School for the Arts: Performance Based Graduation System for the Class of 2008 and Beyond* (Revision: June 28, 2006). Woonsocket, RI: Beacon Charter High School for the Arts.

Schaller, George B. 1963. *The Mountain Gorilla: Ecology and Behavior.* Chicago: University of Chicago Press.

Seligman, Martin E. P., and Steven F. Maier. 1967. "Failure to Escape Traumatic Shock." *Journal of Experimental Psychology* 74(1): 1–9.

Silver, Edward A., and Wendy Metzger. 1989. "Aesthetic Influences on Expert Mathematical Problem Solving." In *Affect and Mathematical Problem Solving: A New Perspective.* Douglas B. McLeod, and Verna M. Adams, Eds. 59–74. New York: Springer-Verlag.

Sinclair, Nathalie. 2004. "The Roles of the Aesthetic in Mathematical Inquiry." *Mathematical Thinking and Learning* 6(3): 261–284.

Snow, C. P. 1990. "The Two Cultures." *Leonardo* 23(2/3): 169–173.

Strauss, Anselm, Leonard Schatzman, Rue Bucher, Danuta Ehrlich, and Melvin Sabshin. 1969. "The Process of Field Work." In *Issues in Participant Observation: A Text and Reader.* George J. McCall, and J. L. Simmons, Eds. 24–27. Reading, Massachusetts: Addison-Wesley.

Swanson, Christopher, B. 2009. *Cities in Crisis 2009: Closing the Graduation Gap.* Bethesda, MD: Editorial Projects in Education, Inc.

Vatterott, Catherine. 2007. *Becoming a Middle Level Teacher: Student focused teaching of early adolescents.* New York, NY: McGraw-Hill.

Warm, Joel S., William N. Dember, and Raja Parasuraman. 1991. "Effects of Olfactory Stimulation on Performance and Stress in a Visual Sustained Attention Task." *Journal of the Society of Cosmetic Chemists* 42 (May/June): 194–210.

Wilbraham, Antony C., Stanley, Dennis D., Matta, Michael S., and Edward L. Waterman. 2008. *Chemistry.* Boston: Prentice Hall.

Wright, S. Paul, Sandra P. Horn, and William L. Sanders. 1997. "Teacher and Classroom Context Effects on Student Achievement: Implications for Teacher Evaluations." *Journal of Personnel Evaluation in Education* 11: 57–67.

DISSERTATIONS

Bross, April J. "An Evaluation of Meaningful Learning in a High School Chemistry Course," (Ph.D. diss., Temple University, 2008), accessed February 3, 2011, ProQuest Dissertations and Theses (AAT 3320003).

Butler, John D. 2009. *Exploring some Inattended Affective Factors in Performing Nonroutine Mathematical Tasks*. Ph.D. dissertation, University of Rhode Island and Rhode Island College. Ann Arbor: ProQuest/UMI (Publication No. AAT 3401142).

JOURNALS

JCE (Journal of Chemical Education). 2001. "Just Breathe: The Oxygen Content of Air" (Classroom Activity: #35). *Journal of Chemical Education* 78(4): 512A–512B.

GOVERNMENT DOCUMENTS

107th Congress. 2002. *No Child Left Behind Act of 2001* (Public Law 107–110—Jan. 8, 2002; 115 Stat. 1425). Washington, DC: United States 107th Congress.

NAS (National Academies of Science). (1996). *National Science Education Standards*. Washington, DC: National Academy Press.

NRC (National Research Council). 2000. *Inquiry and the National Science Education Standards: A Guide for Teaching and Learning*. Washington, DC: National Academy Press.

NRC (National Research Council). 2011. *Successful K–12 STEM Education: Identifying Effective Approaches in Science, Technology, Engineering, and Mathematics*. Washington, DC: National Academy Press.

NRC (National Research Council). 2012. *Framework for K–12 Science Education: Practices, Crosscutting, Concepts, and Core Ideas*. Washington, DC: National Academy Press.

OECD (Organisation for Economic Co-Operation and Development). 2010. "PISA 2009 Results: Executive Summary."

RIDE (Rhode Island Department of Elementary and Secondary Education) 2011a. "Rhode Island's NECAP Science Results for Grades 4, 8, and 11 May 2011 Test Administration (2011)." Providence, RI: RIDE.

———. 2011b. *The Rhode Island Model: Teacher Guidebook 2011-2012*. Providence, RI: RIDE.

INTERNET WEBSITES

Begley, Sharon 2010. "Arsenic and New Life—or Not: Claims of a new life form get whacked." Newsweek. Accessed February 21, 2013. http://www.sharonlbegley.com/claims-of-a-new-life-form-get-whacked.

Chun-xuan, Jiang. 1998. On the limit for the periodic table of the elements. *Aperion* Vol.5 Nr. 1–2, January–April. Accessed December 8, 2012. http://redshift.vif.com/JournalFiles/Pre2001/V05NO1PDF/v05n1jia.pdf.

Dowden, Bradley. 2011. "Time." *Internet Encyclopedia of Philosophy*. Accessed December 8. http://www.iep.utm.edu/time/. Original work published August 11, 2001.

Evertson, Carolyn M. 2003. "Classroom Management." Encyclopedia of Education. *Encyclopedia.com*. Accessed December 8, 2012. http://www.encyclopedia.com/doc/1G2-3403200113.html.

Huitt, William G. 1997. "Classroom Management: First Week Activities." *Educational Psychology Interactive*. Accessed December 8, 2012. http://www.edpsycinteractive.org/topics/manage/firstwk.html.

Lendino, Jamie. 2010. "NASA discovers new life form on Earth." PC Magazine. Accessed February 10, 2011. http://www.extremetech.com/extreme/84010-nasa-discovers-new-life-form-on-earth.

Marsh, Jenni. 2012. "Women do have Higher IQ than Men (Thanks to Multi-Tasking)." *EXPRESS.co.uk*. Accessed December 8, 2012. http://www.express.co.uk/posts/view/333129/Women-do-have-higher-IQ-than-men-thanks-to-multi-tasking.

Office of the Commissioner. 2012. "Charter Schools: Purpose of Charter Schools." Rhode Island Department of Elementary and Secondary Education. Accessed December 8, 2012. http://www.ride.ri.gov/commissioner/charterschools/purpose.aspx.

SADD (Students Against Destructive Decisions) 2012. "SADD's Mission." SADD. Accessed December 8, 2012. http://sadd.org/mission.htm.

REPORTS AND TECHNICAL PAPERS

Bridgeland, John M., John J. DiIulio, and Karen Burke Morison. 2006. "The Silent Epidemic: Perspectives of High School Dropouts" (A report by Civic Enterprises in association with Peter D. Hart Research Associates for the Bill & Melinda Gates Foundation). Washington, D.C.: Civic Enterprises.

Butler, John D. 2011a. "Assessing Charter School Performance: An Exploration of Various Metrics" (White paper, January, 2011). Woonsocket, RI: Beacon Charter High School.

———. 2011b. "Classroom Management: Best Practices" (Technical Brief No. BCHS-TR-080811-Rev.2–081211). Woonsocket, RI: Beacon Charter High School.

———. 2011c. "Comparison of Beacon Charter H.S. NECAP Science Performance with Sending Districts and the State Across Testing Years 2008–2011" (Technical Brief No. BCHS-TR-101211). Woonsocket, RI: Beacon Charter High School.

———. 2011e. "Homework: Foundational Research for Beacon's Guiding Principles" (White paper, March, 2011). Woonsocket, RI: Beacon Charter High School.

NEWSPAPER ARTICLES

Layton, Lyndsey. 2013. "Gates Foundation study: We've figured out what makes a good teacher." *The Washington Post* (January 8, 2013).

Nadeau, Joseph B. 2005. "Beacon to shine on opening day." *The Call* (September 10, 2005). Woonsocket, Rhode Island.

About the Author

John D. Butler, PhD was a member of the Beacon Charter High School faculty from 2005–2013 serving as Mathematics Department Chair and Data Analyst. In 2013, he became the Director of Academic Logistics and Planning for the Village Green Virtual Public Charter School in Providence, Rhode Island.

Born and raised in Warwick, Rhode Island, Dr. Butler earned an AS in Engineering from the Community College of Rhode Island in 1977, followed by a BS in Mechanical Engineering from the University of Western New England in 1979. As a licensed Professional Engineer (PE), he spent twenty-two years working in the field of mechanical/electrical engineering where he designed automated machines and supervised their installation, both in the US and Europe. He is the inventor of over 15 US and foreign patents and the author of several published technical articles in the field of engineering. In 1982, he co-founded the Mechtrix Corporation (a leading manufacturer in the electrical wire processing industry) located in Menomonee Falls, Wisconsin.

After relocating back to Rhode Island in 2001, he began teaching high school mathematics. He earned a Master of Arts in Teaching Mathematics (MAT) from Providence College in 2003 and a Doctor of Philosophy (PhD) in Education from the University of Rhode Island and Rhode Island College in 2009. He is the recipient of the Kappa Delta Pi "2010 Outstanding Doctoral Dissertation Award," from the University of Rhode Island for his dissertation, entitled *Exploring Some Inattended Affective Factors in Performing Nonroutine Mathematical Tasks*. Dr. Butler has also been an adjunct faculty member of the Rhode Island College Upward Bound Program since 2007.

About Mrs. Waters

Jessica M. Waters has been teaching science at Beacon Charter High School (Woonsocket, Rhode Island) since 2008. Since that time, and with her as lead science teacher, Beacon's NECAP (New England Common Assessment Program) science performance has risen from 5% proficient or above to 45%—the second highest gain in science scores of all Rhode Island high schools over the same four-year period. She received the Rhode Island 2013 Teacher of the Year award and was also given the Peter McWalters Professional Educator Award in 2012.

She currently teaches 11th grade chemistry and 12th grade human anatomy and physiology. She holds a degree in Secondary Education, Biology and Chemistry from Rhode Island College. Jessica is a member of the Amgen Bruce Wallace Biotechnology Lab Program and RISTA (Rhode Island Science Teacher Association). She is the participating teacher in the SADD (Students Against Destructive Decisions) after school program. She is happily married and the proud mother of three girls and successfully balances career and family.